JN441622

CONSERVATION AND MANAGEMENT OF THE WORLD'S PETROGLYPH SITES

About the Authors

Hotae Jeon

is a historian specializing in ancient history of Korea. He received his Ph.D. from Seoul National University and is currently professor at the Department of History and culture and director of the Bangudae Petroglyphs Institute, University of Ulsan. He has published numerous books and articles on Goguryeo tomb paintings as well as ancient Chinese art and culture, and has organized many exhibitions on Goguryeo tomb paintings both in and outside Korea.

Kwon-woong Lim

studied conservation and restoration at the Cologne University of Applied Science, Germany, where he obtained his German Diplom-Restaurator. He is specializing in conservation and restoration of stone object and wall painting. He participated in Conservation Project of Wall Painting in Roman Tomb, Nehren in Germany and Angkor Wat Conservation Project in Cambodia, in Project of Cologne Cathedral Conservation. He is participating in ROK/UNESCO Funds-in-Trust Project, "Preservation of Cultural Heritage" in the DPRK, and in several other projects for the inscription of Korean cultural heritage on the UNESCO World Heritage List as a consultant for World Heritage Centre, UNESCO.

Young-hee Park

is a prehistorian and formerly professor at Dongseo University. She graduated from Ewha University (B.A. and M.A. in history) and received a Ph.D. degree from Museum National d'Histoire Naturelle, France. She served as a curator at Dankook University Museum and also had taught at Ewha University and Yeonsei University.

Mounir Bouchenaki

is presently Director of the UNESCO Category II Arab Regional Centre for World Heritage based in Bahrain. Born in Algeria, he obtained his Ph.D. at Aix-en-Provence University (France) and pursued a specialization course in archaeology of Punic and Roman period at Rome La Sapienza University (Italy). He has been in charge of the Cultural Heritage Administration of Algeria from 1972 up to 1982. Since 1982 up to early 2006 he was successively Chief of Section, Director of the Division of Cultural Heritage and Director of the World Heritage Centre, and finally Assistant Director General for Culture. From 2006 to 2012, he served as Director general of ICCROM.

Knut Helskog

is a professor of Archaeology at Tromsø University Museum, UiT the Arctic University of Norway. His responsibilities focus on the management of prehistoric cultural heritage and research on rock art and subsistence-settlement patterns among hunter-gatherer-fisher populations in northernmost Europe.

Ekaterina Devlet

is a professor of Scientific Institute of Archaeology, Russian Academy of Sciences and Director of the Rock Art Research Expedition of the Institute of Archaeology, Russian Academy of Sciences. Her current field projects are related to rock art sites of the Far East region. She has published about 180 papers concerning mainly with open-air rock art studies and preservation and seven books on rock arts in Russia, UK, USA, etc.

Johannes H. N. Loubser

is the owner of Stratum Unlimited, LLC in Atlanta, United States of America, where he specializes in archaeology and rock art. He has worked at rock art sites in various areas across the United States of America and Canada, the Caribbean Islands, Mexico, Bolivia, southern and eastern Africa, and Hawaii. He is a research associate at the University of the Witwatersrand's Rock Art Research Institute and at the Lamar Institute in Georgia.

María de la Luz Gutiérrez Martínez

is an archaeologist researcher at the National Institute of Anthropology and History in Baja California Sur, Mexico, since 1991. She has dedicated her professional life to the study of the archeology of the central part of the Baja California peninsula, and in particular, to the research of its rock art. She has also developed methodologies around the management of this fragile cultural expression. Since 1995 she is responsible for the management of "Rock Paintings of the Sierra de San Francisco." This property was declared a World Heritage Site by UNESCO in 1993.

Jiyeon Kim

is an art historian specializing in Korean painting and visual culture. She received her Ph.D. from the University of California, Los Angeles and is currently research professor of the Bangudae Petroglyphs Institute, University of Ulsan. She also teaches art history and museum studies at the University of Ulsan.

WORLD PETROGLYPHS RESEARCH · II

CONSERVATION AND MANAGEMENT OF THE WORLD'S PETROGLYPH SITES

BANGUDAE PETROGLYPHS INSTITUTE,
UNIVERSITY OF ULSAN

WORLD PETROGLYPHS RESEARCH II
CONSERVATION AND MANAGEMENT OF THE WORLD'S PETROGLYPH SITES

Bangudae Petroglyphs Institute, University of Ulsan
Bldg. 8 Room 111-1 University of Ulsan, Daehak-ro 93, Namgu,
Ulsan Metropolitan City, Republic of Korea 680-749

First edition, 2014
by Hollym International Corp., USA
Phone 908 353 1655 **Fax** 908 353 0255
http://www.hollym.com **e-Mail** contact@hollym.com

Published simultaneously in Korea
by Hollym Corp., Publishers, Seoul, Korea
Phone +82 2 735 7553-4 **Fax** +82 2 730 5149
http://www.hollym.co.kr **e-Mail** info@hollym.co.kr

ISBN: 978-1-56591-140-6
Library of Congress Control Number: 2014905333

Printed in Korea

On Publishing the Second Issue of *World Petroglyphs Research*

The World Petroglyphs Research series was launched in 2013 with the purpose of locating the Bangudae and Cheonjeon-ri petroglyphs of Ulsan in the larger context of the world's prehistoric rock art. Since their discovery several decades ago, the Bangudae and Cheonjeon-ri petroglyphs have attracted huge attention in Korea as well as outside the country. The scale of the site, the variety and uniqueness of the subjects, and the maturity and completeness of the compositions has been well recognized by major international rock art specialists. The significance of these sites has been increasingly highlighted as the Korean government is striving to have them included in the UNESCO World Heritage List and they have been included on the World Heritage Tentative List since 2010. However, a more systematic scholarly study of the sites needs to be undertaken and more active communication with international rock art experts is imperative in order to better understand and promote the value of the two sites.

Last year, the Bangudae Petroglyphs Institute published the first issue of *World Petroglyphs Research*. This year, the institute is publishing a second issue, focusing on the conservation and management of rock art. The preservation of rock art sites, especially of the Bangudae site, has been a huge issue recently. The Bangudae petroglyphs were discovered in 1971, six years after the Sayeon Dam was built. At the time, the petroglyph panel was already underwater from four to seven months a year, causing serious damage to the petroglyphs. For the preservation of the site, the Cultural Heritage Administration of Korea and local environmentalists advocated lowering the water level but the Ulsan Metropolitan City objected to this plan because of concerns that it would result in a shortage of water. The Ulsan city government instead suggested changing the waterway and building a water-blocking wall, but the Cultural Heritage Administration rejected the plan for the reason that it could damage the surrounding natural scenery and the ecological system. Most recently, a plan to implement a transparent kinetic dam around the Bangudae petroglyphs is being considered as a way to protect them without altering the natural environment around them. Currently, the Bangudae and Cheonjeon-ri sites are on the World Heritage Tentative List under the category of culture only, although originally a plan to submit them for both cultural and natural criteria was considered. Preservation of the surrounding landscape, however, is still a major concern for a number of interested parties and a key point of debate.

The chapters in this volume are expected to be read as useful case studies of how other rock art sites facing similar preservation issues are managed and preserved. Five foreign experts specializing in rock art conservation and management of cultural heritage sites contributed articles to this volume. In addition, four Korean scholars wrote about various issues related to management and preservation.

I would like to express my sincerest thanks to the Foundation for Industrial Cooperation of the University of Ulsan and the University of Ulsan for their constant encouragement of the project and their assistance in publishing this book. Special thanks go to Professor Jiyeon Kim and the research staff of the Bangudae Petroglyphs Institute for their commitment to this project and to the Hollym Publishing Company for sharing our enthusiasm in the project.

Hotae Jeon

Director of Bangudae Petroglyphs Institute, University of Ulsan

January, 2014

Contents

01

Preservation of the Bangudae Petroglyphs in Ulsan, Korea

Hotae Jeon

Environmental Transitions around Bangudae

In 1965, with the construction of Sayeon Dam in Ulsan, the Bangudae petroglyph site was submerged, along with several neighboring villages, including Keunmasil, Geonneondeul, Seodangmasil, and Jitongmasil. The site was discovered in 1970, despite the fact that it was only exposed during periods of severe drought. Although the site was designated as Korean National Treasure 285 in 1995, to this day it remains under water for most of the year.

In 2000, University of Ulsan Museum surveyed the site and determined that immediate action must be taken to preserve the Bangudae petroglyphs (Figs 1, 2). Since their discovery 40 years ago, the condition of the petroglyphs has markedly deteriorated, as the carvings are getting smoother and the rock surfaces are chipping away (Jeon H. 1996). This article summarizes the present state of the Bangudae petroglyphs, based on the report by the University of Ulsan Museum (2000), and also considers some possible methods of preservation. In Korea, there are currently around 28 different sites where petroglyphs are exposed to erosive elements and in need of proper preservation (Map 1). The conditions at these sites require serious attention and action from researchers and related institutions.

Fig. 1. A complete view of the Bangudae Petroglyphs

The Bangudae petroglyphs are located in Daegok-ri, which was originally divided into Daegok-ri of Oenam-myeon in Gyeongju City and Daegok-ri of Jungbuk-myeon in Eonyang County. In 1914, during the Japanese colonial era, administrative districts were re-organized, and the two divisions were integrated into Daegok-ri of Eonyang-myeon. In 1965, the Sayeon Dam was built in the Bangudae area, leading to major geographical changes (Maps 2, 3). A considerable number of villages and traffic routes were submerged, including Keunmasil, the largest village in Daegok-ri; the Sin-ri Villages, known as Geonneogakdan or Geonneondeul; Seodangmaeul, a village with a school; and Jitongmaeul, which was famous for paper-making. In addition, Geonneogakdan, the route connecting the villages to Bangu Elementary School, and the valley route along the Bangudae site are no longer traversable.

The basin area of Sayeon Dam is 124.50 km^2, and the dam's storage volume is 2.5 billion m^3. At full capacity, the water reaches up to 63.2 meters above sea level, with an average level of around 60 meters and a maximum height of 66.4 meters during the rainy season of summer. The upper section of the Bangudae petroglyphs is at 55.2 meters, which means that they are usually under water for more than eight months out of the year, depending on how long the winter

Fig. 2. Actual measurements of individual petroglyphs of Bangudae

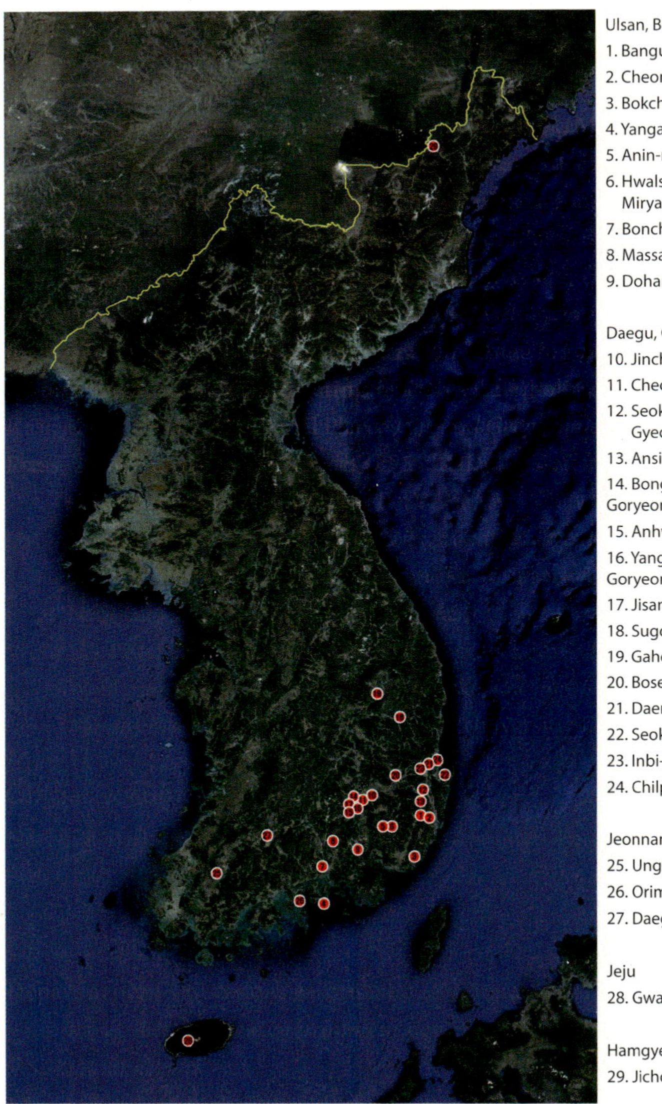

Ulsan, Busan, Gyeongnam
1. Bangudae, Ulsan
2. Cheonjeon-ri, Ulsan
3. Bokcheon-dong, Busan
4. Yanga-ri, Namhae
5. Anin-ri, Miryang
6. Hwalseong-dong, Miryang
7. Bonchon-ri, Sacheon
8. Massang-ri, Uireong
9. Dohang-ri, Haman

Daegu, Gyeongbuk
10. Jincheon-dong, Daegu
11. Cheonnae-ri, Daegu
12. Seokjang-dong, Gyeongju
13. Ansim-ri, Gyeongju
14. Bongpyeong-ri, Goryeong
15. Anhwa-ri, Goryeong
16. Yangjeon-dong, Goryeong
17. Jisan-dong, Goryeong
18. Sugok-ri, Andong
19. Gaheung-dong, Yeongju
20. Boseong-ri, Yeongcheon
21. Daeryeon-ri, Pohang
22. Seok-ri, Pohang
23. Inbi-ri, Pohang
24. Chilpo-ri, Pohang

Jeonnam, Jeonbuk
25. Ungok-dong, Naju
26. Orim-dong, Yeosu
27. Daegok-ri, Namwon

Jeju
28. Gwangnyeong-ri, Jeju

Hamgyeongbuk
29. Jicho-ri, Musan

Map 1. The petroglyph sites in Korea

Map 2. The location of the Bangudae Petroglyphs in Ulsan by satellite picture

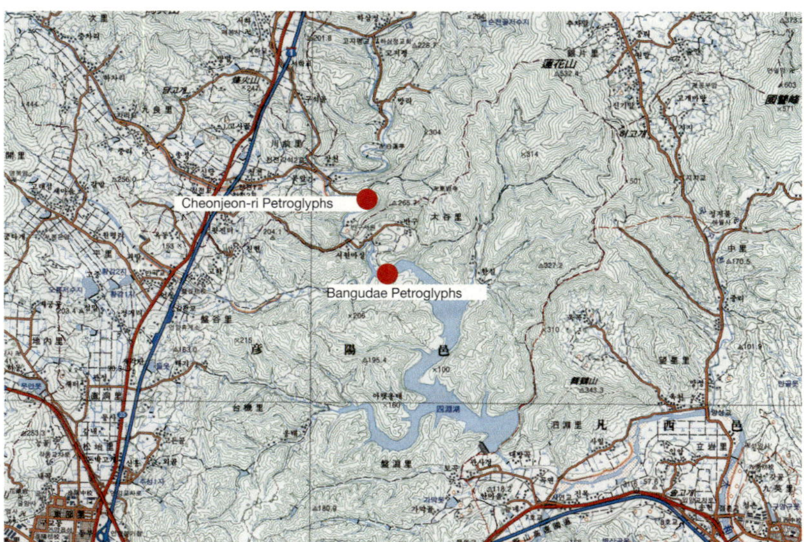

Map 3. The location of the Bangudae Petroglyphs and Cheonjeonri-gakseok Petroglyphs in Ulsan

dry season lasts. Once or perhaps twice per decade, there is an extended dry season, and the panels will remain out of the water for six months or so. But on average, more than 80% of the petroglyphs are submerged for eight to nine months every year.

Such extreme environmental changes clearly have a major effect on the condition of the petroglyphs. The rocks of the Ulsan region generally consist of sedimentary rocks from the Silla Series of the Gyeongsan System (Daegu Formation); their intrusive or eruptive igneous rocks; and the terrain of the Bulguksa Series, formed through the intrusion of multiple Silla rocks (Yi Y. and Yi I. 1972). The lowest part of the Daegu Formation, which is widely distributed in the area around the eastern part of the Eonyang Fault, differs from the upper sheet. The lower sheet consists primarily of dark green, gray-green, and dark gray sandstones, siltstones, sandy siltstones, sandy shale, and shale. The rock layers of the Daegok Stream, on the other hand, mostly comprise violet sandy siltstones, siltstones, sandy shale, and shale. Thus, the two sheets form an alternation of strata, with a bed of dark grey and dark gray-green sandstone, sandy siltstone, siltstone, sandy shale, and shale, with thin layers of dark gray mudstones between the respective layers. The western Eonyang Fault consists mostly of igneous rocks, including andecite, dacite, and porphyry rocks from the Silla Series of the Gyeongsang System. Also, igneous rocks of the Bulguksa Series intrude in the aforementioned sedimentary and volcanic rocks. The igneous rocks of the Bulguksa Series are primarily made up of granite (Eonyang granite and biotite granite) and granitic diorite, porphyry, and dike rocks. The base rocks in this area are unconformably covered by a Quaternary alluvial layer composed of gravel, sand, clay, etc., which is distributed along the sides of tributary streams in the Ulsan area.

The severe erosion and chipping of the panels of the Bangudae petroglyphs is hardly surprising, given that they are rendered on the base stratum of highly hydrated and non-durable sedimentary rocks, consisting of fragile sandstone, sandy siltstone, siltstone, sandy shale, and shale. As is widely known, the major elements of sandstone are quartz, feldspar, and calcite, the latter of which tends to react with water and becomes soluble. This is a major problem for the Bangudae petroglyphs, since calcite is one of the major elements of the rocks in the Daegok-ri area. With their surfaces of dark brown siltstone and composition of albite, quartz, chlorite, orthoclase, calcite, illite, and biotite, the Bangudae panels are highly susceptible to severe water damage.

Whenever the water level drops and the rocks are re-exposed to the air, their surface becomes bleached as a result of a chemical reaction. Specifically, when the water that had been saturating the rocks drains away, it reacts with the calcium in the rocks, causing a chemical decomposition that whitens the rock surface (Yi S. 1994). According to an actual survey conducted by University of Ulsan Museum, this phenomenon has been observed in several rock surfaces at the Bangudae site.

In addition to this chemical decomposition, conditions at the site are continually worsening due to various types of physical, chemical, and biological weathering and erosion. Physically, the rock layers around Daegok-ri are strewn with small cracks and joints that are steadily widening due to the continuous submersion, ultimately resulting in chippage. The rock quality of the porous sedimentary rocks is rapidly deteriorating as a result of the constant absorption of significant fluids. Finally, during any typhoon, the rock walls are abraded by massive rain and pebbles striking the surface. All of these elements facilitate breakage of the rock joints and adversely affect the engravings. The investigation in 2000 confirmed the formation of new fissures, widening of gaps, and abrasion or even effacement of rock engravings (UUM 2000).

The chemical reactions causing the petroglyphs to deteriorate are being intensified and accelerated by various biological elements. In the dry season, when the rocks are exposed, moss grows on them, causing immediate damage from the invasive roots. Moreover, the moss causes further damage when it eventually decomposes, or else falls off, taking fragments of rock with it. The roots of other types of plants also extend down from the upper part of the rock carvings, widening the cracks and expediting the breakage of rock joints. The surface of the rocks is further deteriorated by acid rain trickling down from the upper part of the rock, and by water polluted by fishing or cooking.

Before conducting its survey, University of Ulsan Museum cleaned the area on and around the rocks of environmental hazards, including moss, mud, twigs, bits of rock, and various other waste and deposits around the stone wall. They also identified areas most in need of attention due to severe damage from weathering and erosion. The area with the most significant problems is the southern section of surface A, where the cracks are so severe that the rock panel could seemingly break apart at any moment (Fig. 3). Other areas where the petroglyphs were thinning or being destroyed by chippage included C-88 (Fig. 4) and Rock I (Fig. 5).

Fig. 3. Rock A of the Bangudae Petroglyphs

Fig. 4. C-88 area of the Bangudae Petroglyphs

Fig. 5. Rock I of the Bangudae Petroglyphs

In addition to these environmental hazards, the petroglyphs have also suffered damage caused by people. While the rocks are exposed, visitors sometimes make casts of the petroglyphs, leaving behind synthetic resins on parts of the rock surfaces. Notably, the presence of such chemical resin will cause those parts of the rock to react differently to the water and other natural elements, thereby altering the degree of weathering that occurs. Visitors have also caused damage by making unauthorized stone rubbings of the petroglyphs and by scribbling and scratching on the surface of the rock.

In area C-15, the dashed line in the center of the figure has been badly eroded due to repeated stone rubbings and washings (Fig. 6). The figures of C-18 have also been effaced, such that the outlines of the figure are difficult to identify (Fig. 7). The figures in and around areas C-55 and C-70 are impossible to classify due to surface damage caused by stone rubbing and mold casting (Figs. 8, 9) (Jeon H. 1996). Similar damage can also be seen in areas D-41, D-42, and D-43 (Fig. 10); in particular, the etchings in D-42 are now difficult to discern with the naked eye.

Other cases of chippage, breakage, or other damage caused by mold

Fig. 6. C-15 area of the Bangudae Petroglyphs

Fig. 7. C-18 area of the Bangudae Petroglyphs

Fig. 8. C-55 area of the Bangudae Petroglyphs

Fig. 9. C-70 area of the Bangudae Petroglyphs

Fig. 10. D-41, 42, 43 areas of the Bangudae Petroglyphs

casting can be seen in areas B-19, B-30, D-27 and surrounding pictures. At the time of investigation, B-19 was coated with residue left by prior moldings, and traces of molding were also found on B-30, where the engraved face had been substantially eroded (Fig. 11). Residue also remained on the outlines and stripes of D-27 (Fig. 12), while F-1 had become discolored by ink sediment from excessive stone rubbings (Fig. 13).

Thirty years of weathering, erosion, and artificial damage have taken a heavy toll on the petroglyphs, as seen by comparing the photographs in the University of Ulsan Museum catalogue (2000) to those from a 1984 report, which is based on the first investigation of the petroglyphs in 1970 (Hwang S. and Mun M. 1984). What can be done to prevent further deterioration of the petroglyphs?

Strategies for Preserving the Petroglyphs

The safest and most reliable way of preserving most sites is to attempt to recreate the conditions that existed prior to their discovery and investigation, and the same is true of the Bangudae site. Ideally, this would mean returning to the conditions that existed before the petroglyphs were submerged. However, it is not possible to dismantle the Sayeon Dam, which is now an essential source of water for the city of Ulsan. In this situation, ideal environmental conditions for preserving the petroglyphs may never be achieved.

In that case, what is the best alternative for preserving the petroglyphs? To begin with, the environment must be improved by eliminating or alleviating the most harmful influences. As discussed, the primary cause of damage is the repeated cycle of submergence and exposure of the panels. Therefore, the submergence should be prevented or, at the very least, its duration should be minimized.

There are three main strategies for addressing the submergence of the petroglyphs without significantly affecting the water supply to the city: minimizing the quantity of water flowing to the site; altering the waterway of Daegok Stream; and regulating the water quantity of Daegok Dam, in the upper part of Daegok Stream. Ideally, these three plans can be combined. Adjusting the water quantity of Sayeon Dam and Daegok Dam and regulating

Fig. 11. B-19, B-30 area of the Bangudae Petroglyphs

Fig. 12. D-27 of the Bangudae Petroglyphs

Fig. 13. F-1 of the Bangudae Petroglyphs

the amount of water that flows to the site could prevent the petroglyphs from being submerged, while also helping to control visitor access to the site.

Another possible strategy for mitigating environmental and human hazards is to enclose the petroglyphs in waterproof walls of some type (Kim S. 2000). However, such measures would prevent people from being able to see the petroglyphs within their surrounding environment. The production of petroglyphs is usually strongly linked to the natural environment surrounding the site, and Bangudae is no exception. With its huge cliff set deep inside the valley and near the water, the site seems to have been specially chosen as a desirable setting for the engravings. In fact, the preservation of the petroglyphs for thousands of years prior to human intervention may have been aided by the geographical conditions and climate of the site, such as the angles of the cliff, the amount of wind and sun, and changes in the temperature and composition of Daegok Stream. The construction of walls or barriers would certainly influence these factors in unforeseeable ways, possibly resulting in further damage. Still, if regulating the water quantity of the dams proves to be impossible or insufficient, the construction of protective walls—with the utmost care and consideration—must be considered as an alternative or complementary measure.

Another suggested strategy is the direct application of materials to harden or waterproof the rock itself, or fill the cracks (Crosby A. 1980). This method can only be considered if the safety of the treatment can be absolutely guaranteed. The hardening or joining of the rock surface would cause various physical and chemical reactions that must be carefully evaluated. While there have been multiple reports of similar methods used to preserve relics, any direct comparisons are problematic given the abundant differences in the characteristics and materials of the relics. With petroglyphs, for instance, one must consider the difference between carved and non-carved cases, severe and slight erosion, submersion and non-submersion, cracked and non-cracked surfaces, etc.

In addition to these direct methods, some indirect methods could also be used to enhance preservation to varying degrees. First and foremost, all visitors and researchers must be adequately informed about the delicate nature of the site, and urged to refrain from any direct contact with the petroglyphs. Legal measures, such as having the Bangudae petroglyphs, the

Cheonjeon-ri petroglyphs, and the surrounding area designated as a protected site, should be explored. Such designation would require a comprehensive academic investigation of the Daegok Stream area, which would ameliorate the restoration of the site's conditions prior to its discovery.

Concurrently, a life-size replica of the petroglyphs should be produced using a proven technology that conforms to international standards for cultural property protection and reproduction. Given the difficulty of examining the petroglyphs from any distance, and the damage that results from direct contact, a replica is necessary to serve the needs of researchers and others who wish to observe the petroglyphs up close. Multiple replicas could be produced and installed in places that are open to the public, including an exhibition hall with a research center dedicated to the petroglyphs. For improved access and adequate space for facilities, it might be best to locate such a building outside the immediate area of Bangudae, perhaps somewhere near National Road Number 37. The related construction of any buildings or roads must avoid the 2.6 kilometre gorge entering Bangudae for conservation reasons.

Along with the production and display of life-size replicas, a proper observation system should be established to meet the interests and demands of researchers and the public. Above all, it is essential for observers to maintain a certain distance from the petroglyphs. Thus, signs with informative contents should be installed on the hill opposite of Daegok Stream, and a small observatory with telescopes should be constructed for visitors' convenience. Of course, these artificial facilities must be kept as minimal as possible.

The Bangudae petroglyphs and the Cheonjeon-ri petroglyphs have been a crucial catalyst for research on Korean petroglyphs, attracting major attention from related researchers. Despite this interest, and the designation of the site as a Korean National Treasure, the area remains largely unsupervised and unprotected. It is little wonder, then, that the condition of the petroglyphs has continuously deteriorated for the past thirty years. To stop this trend, appropriate supervision and preservation plans must be implemented. Hopefully, through the support of academia and related institutions, the ongoing cycle of submersion and exposure that is causing the severe erosion of the Bangudae petroglyphs can be ceased as soon as possible.

BIBLIOGRAPHY

Crosby, A. 1980. "Conservation of Painted Lime Plaster on Mud Brick Walls at Tumacacori National Monument, U.S.A.," conference paper for Third International Symposium on Mudbrick (adobe) Preservation, Ankara ICOMO, ICOMOS.

Hwang, Suyeong and Mun Myeongdae. 1984. *Bangudae ambyeok jogak,* Dongguk University Press.

Jeon, Hotae. 1996. "Ulju Daegok-ri, Cheonjeon-ri amgakhwa," Korean Association for History and Folklore, eds. *Hanguk ui amgakhwa,* Hangilsa Publishing.

Jeong, Yeonghwa and Yi Chaegyeong. 1994. *Gyeongju munhwa yujeok jipyo josa bogoseo-Geoncheon-eup, Naenam-myeon, Oedong-eup.*

Kim, Sujin. 2000. "Ulsan amgakhwae bojon munjae," University of Ulsan Museum, *Ulsan Bangudae Amgakhwa.*

University of Ulsan Museum. 2000. *Ulsan Bangudae Amgakhwg.*

Yi, Sangheon, 1994. "Cheonjeon-ri gakseok gwa Daegok-ri amgakhwa ui bojon e daehan jeeon," presented at the fifth Symposium of the Liberal Arts Institute, University of Ulsan.

Yi, Yunjong and Yi Ingi. 1972. *Jijil dopok seolmyeongseo (Eonyang dopock) 1:50,000, Gungnip jijil josaso.*

02

The Conservation and Use of the Pohang Chilpo-ri Petroglyphs

Kwon-woong Lim

Introduction

Petroglyphs are found throughout the world and in various cultures. Petroglyphs, along with dolmens, make it possible to study the life ways and mental landscapes of prehistoric communities from a positivist approach. As rock art sites were formed over a long period of time, they also allow us to examine changes in climate and other aspects of the natural environment, and observe developments in prehistoric lifestyles.[1] The Matobo Hills site in Zimbabwe is a key example, as it is possible to observe from the images a transition in subsistence production; the images, which were produced from the Stone Age to the early Historic period, illustrate a transition from a hunter-gathering group to a farming society.[2] On the other hand, changes in the natural environment can be observed in the images of the rock art site Tadrart Acacus in Libya, which dates from 12,000 BCE to 100 CE; visible in the images are the changes that took place in the lifestyle and the fauna and flora of the

[1] For example, the Rock Drawings in Valcamonica, Italy, which were added to the World Heritage List in 1979, consist of images representing various themes (prehistoric symbols, farming, seafaring, war, magic) that were produced from the prehistoric period (around 8,000 BCE) to throughout the Roman, Middle Age and Modern periods. The rock art images of the Chongoni Rock-Art Area in Malawi were produced from the Neolithic period to modern times by hunter-gatherers and Iron Age farmers.

[2] The images of the Kondoa Rock-Art Site in Tanzania (added to the World Heritage List in 2006) deal with the transition from a hunter-gathering way of life to a farming way of life.

Sahara region.

In Korea, the academic research of petroglyphs began with the discovery of the Bangudae and Cheonjeon-ri petroglyphs, both located in the Ulsan area, in 1970 and 1971, respectively, and with the investigation of the Yangjeon-dong petroglyphs, located in Goryeong, in 1971. Since then, the additional discovery of Korean petroglyphs and various studies undertaken on rock art have resulted in the establishment of rock art studies as an important field of research that can contribute greatly to our understanding of prehistoric life ways and mental landscapes. Korean researchers have begun recently to actively interact with the rock art specialists of other East Asian regions and Northern Europe and this has led to a broadening of research interests, as is evidenced by the undertaking of comparative studies with the rock art of these regions. In addition, efforts have been made to go beyond an understanding of the petroglyphs as an object of academic research and present them to the general public as an important part of our cultural heritage.

Since the early 2000s, the preservation of rock art has become an important topic of debate as a result of the ever increasing threat of damage to the Bangudae petroglyphs following the construction of the Sayeon Dam and the subsequent discussions that have taken place regarding the need for their preservation. Consequently, much investigation and research has been carried out in order to establish an outline for the preservation of Korea's petroglyphs, but a concrete methodological approach on conservation practices has yet to be formulated. In addition, much preparation work has been carried out as part of a continuing effort to nominate the Bangudae petroglyphs to the World Heritage List.

The Pohang area is home to many rock art sites, the most representative of which is the Chilpo-ri site. The petroglyphs of Chilpo-ri, along with those found in the Gyeongbuk region (Goryeong, Yeongju, Gyeongju, Yeongcheon), comprise a category of rock art that is referred to as the "Chilpo-ri type"; a different petroglyph type is represented by the Bangudae and Cheonjeon-ri examples found in the Ulsan area. Given that this stylistic difference can be seen to represent a difference in the cultural origins of the communities that produced the rock art, clearly the preservation of the Pohang petroglyphs is extremely meaningful in terms of their research value.

The conservation management of rock art has been addressed by many researchers but these discussions have tended to be abstract or piecemeal in nature. Therefore, this paper aims to present a relatively detailed program for the conservation and use of rock art in Korea, with a focus on the petroglyphs that are distributed in the Chilpo-ri area of Pohang.

Rock Art Heritage: Conservation Management and Dissemination

1) Survey

The petroglyphs of Chilpo-ri and Sinheung-ri in Pohang are mostly distributed among the plains that surround the lower reaches of the Gohyeon River, near the river mouth, and along the mountain ridge (located at approximately 50 meters above sea level) that connects Mt. Gollyun (176 m) and Mt. Obong (178.5 m).

In contrast to other Korean petroglyph sites, the rocks and rock outcrops that contain the Pohang petroglyphs are clustered in a relatively restricted area, with the individual rock panels containing a single image or multiple images. Therefore, in order to obtain the basic data that is required for the establishment of a long-term conservation management plan, a thorough investigation of the number and characteristic features of the Pohang petroglyphs must be carried out.

From the perspective of conservation science, Korean petroglyphs can be classified into the following three types according to the nature of the base rock upon which the images were carved: the rock outcrop type, the individual rock type (such as a dolmen capstone), and the rock face type.[3] This system of categorization, in conjunction with the current method of petroglyph classification used by Korean scholars, can be of great use in establishing conservation management plans or in identifying the characteristic features of each petroglyph scene.

The investigation of the Pohang petroglyphs must begin with the mapping of the exact location of each petroglyph scene using global positioning system (GPS) equipment. In addition, each of the scenes must be given a name that

[3] It should be noted that this method of classifying petroglyphs into the rock outcrop type, individual rock type, and rock face type, according to the nature of the base rock, was first devised by the author.

is agreed upon and commonly used by all researchers. (At present, they are confusingly referred to by various names). The location of the petroglyphs and the type of rock upon which they were carved comprise the basic data set on which a comprehensive conservation management scheme can be established. Until now, this data has only been compiled in a piecemeal fashion by individual researchers. However, in order to implement a comprehensive management system for the petroglyphs, the rock art must be studied and recorded in their entirety.

Table 1. Example of petroglyph classification according to base rock type

Type of base rock	Example
Rock outcrop	Chilpo-ri Zone I Panel 2,[4] Sinheung-ri Zone VII Urine Rock
Individual rock	Chilpo-ri Zone III dolmen capstone, Chilpo-ri Zone IV round pecked stone (altar stone)
Rock face	Bangudae, Cheonjeon-ri

2) Assessment

In the case of the "rock outcrop" type petroglyphs, the currently visible rock panel may in fact represent only a section of the original rock panel. This may be due to geographic changes in the surrounding area which occurred after the images were carved into the rock surface. The fact that previously unknown cup marks have been found on rock outcrops following the removal of weeds and dirt during investigation points to this possibility (Kim 2006:100). The discovery of cup marks forming a *yut* board pattern near the edge of the rock panel of the Sinheung-ri Urine Rock, which were found during the process of clearing up the site, is also indicative of this possibility (Figs. 1, 2). In excavating the areas surrounding rock outcrops which contain rock art images, it may become possible to identify the exact scale of the petroglyphs. Excavations may

[4] The names used in the current paper for the petroglyphs of the Pohang area have been adopted from the survey report provided by KOPRA and City of Pohang (2007). Those petroglyphs not mentioned in this volume were referred to by commonly used names.

Fig. 1. *Yut* board pattern cup marks newly discovered at the Sinheung-ri Urine Rock (left)
Fig. 2. Exposed view of a "rock outcrop" type petroglyph (Chilpo-ri Zone I Panel 2) (right)

also lead to additional studies on newly discovered images.

In the case of the "individual rock" type petroglyphs, there are also instances in which the lower sections of the rock panel have come to be buried in the ground. Even if the petroglyphs themselves are not covered, the earth should be removed for conservation purposes as it may lead to increased humidity or the ascent of water-soluble salts due to capillary action, which could result in damage to the petroglyphs. Therefore, the entire rock panel containing the petroglyphs must be cleared of earth, and the surrounding grass and scrub must be removed in order to lower the humidity around the rock, thereby preventing the growth of lichen or mold.

The most difficult aspect of rock art research is reconstructing prehistoric life ways and mental worlds based on piecemeal images. Rock art images convey only an extremely limited amount of information; in the case of the Chilpo-ri petroglyphs, the majority of the scenes consist of abstract images and cup marks. Therefore, little can be said about the meaning or the stylistic lineage of the rock art images based on iconographic research alone; indeed, such an approach can only provide limited insight into when and by whom the rock art images were produced, or the social contexts of their production. In order to overcome these shortcomings, petroglyph sites must also be studied from an archaeological perspective.

The Chilpo-ri petroglyphs are located in a region that is particularly suitable for farming and fishing. Thus, this landscape context may provide clues regarding the types of subsistence activities that were undertaken by the communities that produced the petroglyphs, as well as where these communities may have lived. However, in order to gain a clearer understanding of the producers of the Chilpo-ri petroglyphs, an archeological research plan that involves the survey and excavation of the surrounding areas must be implemented.

For example, in the case of the Chilpo-ri Zone II cup mark stone or the Chilpo-ri Zone IV round pecked stone (both of which are referred to as "altar stones" since the rock art images have been carved into individually standing rocks), an excavation of the surrounding area may yield crucial evidence that can be of use in interpreting the meaning of the petroglyphs.

Archaeological excavations have been carried out even at rock art localities designated as World Heritage Sites. The results of such excavations have yielded important information on the settlements and artifacts of the region, as well as on when the rock art was produced and the nature of the communities that produced them.[5] A similar approach has been adopted in the study of the petroglyphs of Yangjeon-dong in Goryeong with research being conducted on the pottery and stone tools recovered from the area around the rock panel in order to obtain more information about the rock art (Bak 2002:33-49).

GIS Database Creation

A geographic information system (GIS) can be defined as a system that makes it possible to continuously store, compile, and edit maps and images, and manipulate and merge such data with other types of geographical data. In its early days, GIS was mainly used by the United States Department of Defense in order to analyze and create a database of the nation's key industries, such as the defense sector and the industrial sector. However, in the last decade or so, with the development of computers and digital information, GIS has come to be actively used commercially as well as academically. In Korea, GIS has begun to be utilized (albeit at a basic level) in heritage management.

[5] Investigations carried out at the site of the Petroglyphs within the Archaeological Landscape of Tamgaly in Tanzania and at the site of the Ecosystem and Relict Cultural Landscape of Lopé -Okanda in Gabon are examples of this.

A number of components are required in order to apply GIS to the management of rock art. They include the results of prior research carried out on the rock art, a database which contains mapping and other types of information, and the infrastructure and personnel needed to competently merge map-form geographic data with other types of geographic data. Of these components, the actual rock art data, which should be studied and gathered by researchers of the field, plays a central role in determining the overall quality of the GIS database. The competent merging of this data with other types of geographic data determines the efficiency and usefulness of the database. Therefore, it is when specialists of both fields are able to cooperate in a productive manner that good results can be expected.

In the case of the Chilpo-ri petroglyphs, the results of the previous research (comprising the coordinates of the rock art images, photographs, illustrations, research articles containing discussions and interpretations on the rock art, and additional reading lists) must be efficiently merged with the geographic data (maps) so that the basic data concerning the petroglyphs can be easily searched for and downloaded from the maps.

The process of creating a GIS database provides an opportunity to systematically review and analyze pre-existing data, and to identify areas that require additional research. Therefore, the work carried out as part of this data building process can also be used as a basic data set to be used for the establishment of conservation management plans. It can also act to generate new research. In addition, a GIS database will facilitate the sharing of information with the general public. Easier access to information on the Chilpo-ri petroglyphs will increase public interest in the rock art and lead to greater publicity.

Scientific Conservation

Although it has been more than 40 years since the Korean petroglyphs were first reported officially to the academic community, conservation actions based on investigation and research undertaken by conservation specialists have yet to be carried out on the rock art sites, and the infrastructure and conditions necessary for such actions are sorely lacking. Up until now, the conservation of the Korean petroglyphs has been carried out by non-specialists[6] and the approaches and methodologies they used have been problematic. There have even been instances in which conservation actions have resulted in the deterioration of the rock art. This is indicative of the currently problematic situation of rock art conservation in Korea.

Due to this situation, there has been a tendency among certain sections of the academic community specializing in rock art research and art history to unconditionally oppose all conservation actions, based on the argument that such actions will inevitably lead to the rock surface being damaged. In addition, the news that "conservation actions on rock art at Russian sites actually resulted in the acceleration of the rate of deterioration" has come to be wrongly regarded by some researchers as a formally published research finding. However, before any debate can take place on conservation strategy, work must first be carried out on the following: base rock classification, identification of the climatic conditions of the rock art site and the environmental conditions of the rock panel, and the establishment of basic guidelines for the types of situations requiring conservation actions and the types of conservation treatments that are possible, as well as when and how they are to be applied. In addition, in order to objectively judge whether conservation actions have actually accelerated the deterioration process, a database containing scientific information on key elements before and after conservation treatment is needed in order to make this assessment based on quantifiable data rather than mere observation. Such an assessment must be collectively made by a number of conservation scientists and petrologists.

1) The Conservation Program of the Côa Valley Archaeological Park

Viable strategies for the conservation of petroglyphs will be examined in this section with a focus on the conservation program of the Côa Valley Archaeological Park (Fernandes 2007). This will provide the basis for the establishment of a conservation program for the Chilpo-ri petroglyphs.

[6] As recently as a couple of years ago, conservation practices were undertaken by individuals specializing in the fields of chemistry, art history or architecture, rather than by conservation scientists. As can be expected, the former "specialists" were only knowledgeable about one or two treatments and had gained their expertise through a few weeks of training at mostly foreign institutions. Therefore, the conservation actions they carried out were inevitably problematic; the problem of having to undertake additional conservation treatments at these sites has recently emerged.

The rock art sites of the Côa Valley, in Portugal, were designated a World Heritage Site in 1998. In 2010 this World Heritage property was expanded to include the rock art sites of Siega Verge, in Spain. The petroglyphs of the Côa Valley, which date from around 22,000 to 10,000 BCE, were carved into the vertical schist panels that flank the Côa River, a tributary of the Douro River. Approximately 400 rock art panels containing hundreds of human images and thousands of animal figures have been found throughout an area stretching over 17 kilometers. (Figs. 3, 4)

The Côa Valley rock art conservation program consists of two components: investigative research and conservation actions. Investigative research, carried out from 1996 to 2001, includes the investigation and recording of the rock art, a geological study on the stability of the schist outcrops,[7] a biological study on the lichen colonization that can be observed on the rock art panels and the surrounding schist outcrops, a multi-disciplinary study on the weathering of the rock panels, and a study on the siliceous layer that acts to protect the petroglyphs. From October 2003 to October 2005 a seismic activity monitoring system was put into operation in order to measure seismic activity in the Côa Valley. In 2004 a weather station was installed to monitor daily, weekly, and monthly fluctuations in temperature, precipitation, and relative humidity. Measurements were also taken of the insides of non-engraved rock outcrops.

Fig. 3. The location of the Côa Valley rock art site and other rock art sites in the Iberian Peninsula (Photo: Fernandes 2007:76) (left)

Fig. 4. Bird's eye view of the Côa Valley (Photo: Fernandes 2007:77) (right)

[7] The majority of the petroglyphs of the Côa Valley were carved into rock outcrops.

Conservation treatments carried out on the Côa Valley rock art include the cleaning of the petroglyphs and the consolidation of the areas surrounding them. Rock panels are occasionally cleaned to remove the muddy sediment and rubbish that are left on the carvings following the flooding that occurs at the rock art sites. Water from the Côa River and a sponge are used and negative changes in petroglyph conditions have not been observed since cleaning using this method was first undertaken in the winter of 2000/2001. As for the removal of lichens, there was much controversy on the efficacy of, and possible damage caused by, lichen removal (as is the case in Korea). The clearance of lichens from several rock art panels has taken place in the Côa Valley, and future instances of lichen removal will require a meticulous case by case study of the benefits and harm of such intervention (Fernandes 2007).

The greatest challenge facing the conservation of the Côa Valley rock art is the tendency of the schist rock faces, where the majority of the petroglyph rock panels are located, to self-splinter (Fig. 5). Experiments on the consolidation of the fractured schist outcrops were first carried out in 1995. Based on the results of the study on the geomorphologic features of an un-engraved schist rock sample, epoxy resin and a reinforced Portland based mortar was injected into the existing fractures (Figs. 6, 7). A study of this sample 13 years later revealed that the epoxy material had turned yellowish and opaque while the mortar demonstrated no alteration. Therefore, an additional study using mortar was carried out in 2004 (Fernandes and Rodrigues 2008).

The rock art sites of the Côa Valley have experienced some vandalism in the form of graffiti done with sharp objects such as rocks, keys, and knives on the non-engraved surfaces (Fig. 8). Such graffiti marks can be easily removed. However, the decision to remove graffiti is not always easy since additional markings dating from the Upper Paleolithic, Neolithic, and Iron Age have been superimposed onto the original petroglyphs. At present, the graffiti has been recorded and documented in preparation for possible removal in the future (Fernandes 2007).

2) The Conservation of the Chilpo-ri Petroglyphs

Conservation must be preceded by research and investigation from a multi-disciplinary approach as it is only on the basis of the results of such research and investigation that a conservation program can be established. However, a review of the previous research and investigation carried out on the Chilpo-ri petroglyphs will not be presented here; this will be done at a future opportunity. Rather, the focus of this section will be the conservation actions that have been the most controversial; viable conservation methods will be suggested based on a comparative examination of the Côa Valley examples.

Fig. 5. The joints of the schist outcrops (Photo: Fernandes and Rodrigues 2008:112) (left)
Fig. 6. Detail of vertical gap filled by mortar substance (Photo: Fernandes and Rodrigues 2008:117) (right)

Fig. 7. The filling of a fracture gap with mortar (Photo: Fernandes and Rodrigues 2008:117) (left)
Fig. 8. Graffiti covering an un-engraved surface (Photo: Fernandes 2007:91) (right)

(1) Lichen Removal

The controversies surrounding conservation actions that were noted in the earlier examination of the Côa Valley Archaeological Park conservation program are also relevant to the conservation of the Korean petroglyphs. The removal of lichens, in particular, has been a fiercely debated topic. Those who oppose lichen removal do so for the following reasons: it can bring about damage to the rock surface containing the engravings; the damage resulting from it is greater than the damage caused by lichen growth; and the re-colonization of lichens will take place in 3 to 4 years. These concerns are not based on scientific fact. However, those who champion the removal of lichens have not been able to present an objectively valid argument for lichen removal. This absence of objective guidelines by which the efficacy of lichen removal can be assessed may be regarded as a problematic situation. But given the differences in the base rock upon which the petroglyphs were engraved and the diversity of the surrounding environment, it can be argued that an attempt to establish such a general guideline is in itself unscientific and unrealistic.

The colonization of lichens on rock surfaces can lead to the deterioration and loss of material from the rock surface as lichens produce metabolic byproducts such as organic acids. Therefore, lichen removal at an appropriate time is required. This removal of lichens not only serves conservation purposes but also can act to greatly enhance the readability of the rock art images (Figs. 9, 10). A clearer view of the petroglyphs will not only aid researchers, but also give the general public a better viewing experience.

The decision to remove lichens and the method of doing so must be determined in consideration of the nature of the base rock, the surrounding environment, and the patina covering the engravings. Using chemicals to remove lichens should be avoided as it can damage the rock surface. The decision to remove lichens and the method of removal should not be determined by a government administrator but through discussions between conservation scientists, petrologists, and rock art researchers.

In 2008 and 2009 lichens were removed from the rock panels of Chilpo-ri Zone I and Zone III, and the Sinheung-ri Urine Rock. Continuous monitoring must be carried out on these panels in order to assess the efficacy of the lichen removal. In addition, new methods of lichen removal must be developed and experimented with on un-engraved rock outcrops. In the case of the rock

Fig. 9. The Chilpo-ri Zone I petroglyph before lichen removal (left)
Fig. 10. The Chilpo-ri Zone I petroglyph after lichen removal (right)

panels of Chilpo-ri Zone I, research must also be carried out on the structural properties of the base rock.

(2) Stone Consolidation – Rock Surface

The Chilpo-ri petroglyphs were subjected to several forest fires which damaged the surface of the rock panels. The rock surfaces appear to be in a very fragile state, and it is possible to observe both the flaking of the rock material and the chipping of mineral particles. Therefore, the consolidation of the rock surface is needed urgently.

The exact name and type of material that was used to consolidate rock surfaces in the Côa Valley conservation program could not be identified. But given that it was a silicate-based treatment, it is possible to infer the use of ethyl silicate or methyl silicate.

In undertaking conservation of the Chilpo-ri petroglyphs, a study of the reasons for the flaking and cracking of the rock surface, as well as a petrographic identification of the base rock and a study of its physical properties, must first be carried out before the method of rock surface consolidation can be determined.

Therefore, the petrographic studies that have been carried out on some of the Korean rock art sites must be expanded to include the Pohang petroglyphs. In the case of the rock panels of Chilpo-ri Zone I, weathering has brought about the deterioration of certain sections of the rock panel, with severe

Figs. 11, 12. The Chilpo-ri Zone I petroglyph showing erosion of the rock surface

surface erosion resulting in the formation of a muddy layer. Petrographic studies on the mechanism of erosion and its relationship with the surrounding environment should be carried out as well (Figs. 11, 12).

A prior study of the physical properties of the base rock plays a key role in determining the success or failure of conservation efforts to consolidate the rock surface. As the physical properties of the base rock that are deemed to be important from the perspective of conservation science, as well as the methods of studying those properties, differ from the approach and methods used in the fields of geology, architectural engineering, and petrography, such studies must be undertaken by conservation scientists.

(3) Stone Consolidation – Rock Fracture

Fortunately, the Chilpo-ri rock art panels do not contain many fractures or cracks. However, in the case of the Sinheung-ri Urine Rock, much flaking can be observed from the rock panels, a clear indication of the urgent need for consolidation treatment. In the case of the Côa Valley rock art examples, consolidation experiments using a Portland based mortar were carried out. However, cement reinforced mortar contains a large amount of water-soluble salts which may act to hasten the process of deterioration (Fig. 13).

In the case of rock art sites, the consolidation of stone that has experienced flaking and fracturing must be undertaken using an inorganic binding substance specifically made for that purpose,[8] and which does not contain water-soluble salts and has a hardness which is less than that of the stone being subject to consolidation treatment. In 2008 the writer developed, for the first time in Korea, an inorganic mortar for conservation purposes that did not contain water-soluble salts. This mortar was used in conservation treatments carried out on the Geumgang ordination altar of Tongdosa Temple in Yangsan, and the results of this conservation work have remained in good condition during the past three years (Fig. 14).[9]

The inorganic binding substance that is currently being used in Korea for purposes of conservation at cultural heritage properties contains a large percentage of water-soluble salts and was originally developed for general use in building construction. Therefore, the decision to use it must be undertaken with upmost caution. The conservation scientists working on the rock art of the Côa Valley conducted an experiment to see how the mortar used to consolidate fractured stone might alter over time by exposing a sample to the elements for 13 years. Although this was a necessary experiment in terms of cultural heritage conservation, it is realistically impossible to carry out a similar experiment in the Korean context. Therefore, a more viable option is to produce a sample and carry out a lab-based artificial weathering experiment. If the mortar used in the conservation treatments at the Tongdosa Temple is modified in compliance with the physical properties of the base rock forming the Sinheung-ri Urine Rock or the Mt. Gollyun trapezoid-shaped rock of Chilpo-ri Zone II, it can then be used in consolidation treatments aimed at repairing surface flaking and cracks in the rock (Fig. 15).

[8] A substance that takes into account the density, mass, capillary water absorption coefficient, capillary water absorption, total water absorption rate under atmosphere, total water absorption rate under vacuum, compressive strength, contraction rate etc.

[9] The inorganic binding substance that was used in the dismantlement and conservation of the brick pagoda at Bulyeong Temple, Cheongdo, contained large amounts of water-soluble salts such as calcium magnesium oxide (10-20%) and monocalcium phosphate (10-15%). Therefore, it is inappropriate for use on cultural properties. In addition, it is also mixed with a 20 wt.% non-crystalline aluminosilicate substance, which gives it a composition not dissimilar to cement. The misguided notion that damage due to efflorescence (resulting from the cement) could be mitigated through the use of this binding substance was based on a lack of understanding on the basic composition of the inorganic binding substance and its relationship with water-soluble salts (Cheongdo-gun 2009:108-109).

Fig. 13. Damage caused by water-soluble salts contained in Portland based mortar. (Geumgang ordination altar, Tongdosa Temple, Yangsan)

Fig. 14. The results of conservation using an inorganic binding that does not contain water-soluble salts (Geumgang ordination altar, Tongdosa Temple, Yangsan)

Fig. 15. The state of damage observed at the Mt. Gonryun trapezoid-shaped rock of Chilpo-ri Zone II

(4) The Protection of Rock Art Sites through the Establishment of Visitor Pathways

With the exception of the Chilpo-ri Zone I petroglyphs, none of the rock art sites have been provided with information panels. This not only acts to inhibit the general public's access to the sites but also acts as a detriment to their awareness of the significance of the sites. In addition, the sites have not been provided with even the most basic of facilities needed for their protection. Given this situation, it is not surprising that hikers and residents of the surrounding villages have come to incorporate the engraved surface of the Sinheung-ri Urine Rock into their hiking route. This has resulted in serious and continuous damage to the petroglyphs. In the case of irregularly distributed heritage sites, such as rock art sites, visitor pathways can be used to both enhance accessibility and manage movement. By regulating the distance separating the visitor pathway and the rock art, as well as the height of the pathway from the ground, access to the rock art can be prevented without the use of additional facilities.

(5) Management of the Surrounding Area

The areas surrounding the Chilpo-ri rock art sites have not been well managed, with the exception of Chilpo-ri Zone I. In the case of Chilpo-ri Zone II, the uncontrolled growth of vegetation has made it impossible to access several of the rock panels. Vegetation increases the humidity of the surrounding area, which in turn leads to the active growth of organisms such as lichen. As the metabolic byproducts produced by lichens, such as organic acids, can lead to the deterioration of the rock surface, they must be removed. In addition, in the case of the Sinheung-ri Urine Rock, rainwater channeled into the surrounding areas has resulted in the accumulation of earth with a high moisture content. This has provided an ideal micro-climate for the erosion of the rock surface and the growth and expansion of mold. Rainwater contained within holes in the rock may freeze and bring about damage when temperatures drop suddenly (Figs. 16, 17). Therefore, the creation of drainage systems to divert rainwater away from the rock art must be undertaken. Should rainwater continue to percolate through rock outcrops, consolidation treatments must be carried out, although they may not be as successful due to the presence of rainwater.

Fig. 16. Heavy vegetation makes accessing the petroglyphs of Chilpo-ri Zone II difficult.
Fig. 17. Sinheung-ri Urine Rock rainwater flow path

Promotion and Dissemination

The promotion and dissemination of rock art heritage and its conservation represent two sides of a coin. The most traditional mode of promoting and disseminating information on rock art is through permanent or special exhibitions hosted by museums. The conservation of cultural heritage can only be truly meaningful when such attempts to promote and disseminate it occur. This has recently begun to take place in a more active manner due to the notion of heritage sites as a "resource for tourism." In particular, local authorities have actively nominated the representative heritage sites of their region to the World Heritage List. The utilization of heritage sites as tourist attractions—as long as their integrity remains intact—can play a positive role in allowing the general public to experience the sites for themselves. However, the excessive commercialization of cultural heritage or exhibitions that have been organized mainly in order to fulfill administrative goals must be avoided.

The utilization of rock art sites as tourist sites can provide a good opportunity for the active promotion of the Chilpo-ri petroglyphs to the general public. This may lead to economic opportunities for the local community which can bring about the realization that heritage sites may hold the keys to regeneration and new paradigms of growth rather than merely being a hindrance to local development. Such a change in perception will play a key role in obtaining the cooperation of the local community in protecting the region's cultural heritage.

1) Situating the Rock Art in the Regional Context

The Chilpo-ri petroglyphs are located in a region which, as it also contains a beach, a hilly landscape with low-lying mountains (with altitudes of less than 170 meters above sea level), a river and a seaside fishing village, has the potential to become a tourist attraction. The development of a hiking trail or mountain bike course which begins at Chilpo Beach and continues along the foot of the eastern slope of Mt. Gollyun, passing by the various rock art panels of the area and providing a view of the sea and the mountains, is a viable option. Given that constellations have been identified among the images of the Sinheung-ri Urine Rock petroglyph, it may also be possible to organize a stargazing camp, in which participants observe the stars in the night sky and compare them to the Sinheung-ri petroglyph constellations. This could promote an understanding of rock art as a dynamic part of Korea's cultural heritage that is relevant to the present, and is not just an abstract remnant of the past.

2) Linking the Rock Art to Other Cultural Heritage Sites

Approximately 60 dolmens are located in and around the Chilpo-ri area. In addition, tombs and settlement sites, as well as artifacts dating to the Paleolithic and Bronze Age periods have been found in Heunghae-eup, the administrative district of Pohang to which Chilpo-ri belongs. Chilpo-seong Fortress, which was constructed in the fifth year of the reign of King Jungjong of the Joseon Dynasty (1,510 CE), is also in the area, and it is recorded that twelve fortresses, including Mijilbuseong Fortress, were built in the Heunghae region in the ninth month of the fifth year of the reign of King Jijeung of Silla (504 CE) (Yi 2006). The excavation and systematic management of these sites will make it possible to develop a tourist attraction in the region which connects the prehistoric, Three Kingdoms, and Joseon periods.

In order to fully utilize the Chilpo-ri petroglyphs as a tourism resource, a comprehensive heritage management plan has to be established which not only includes the conservation of rock art but also involves the development of facilities that may allow the public to better access and enjoy the rock art sites. In addition, as numerous forest fires in the region have brought about the destruction of certain parts of the natural landscape in which the rock art sites are distributed, active attempts to restore the surrounding vegetation through the planting of adult trees must be also be included in the comprehensive heritage management plan. This long-term management plan must also consider the need for guard posts and a central management center.

Conclusion

Since the existence of the Chilpo-ri petroglyphs was first reported to the academic community in 1989, they have been studied by researchers from various fields. However, the conservation management of these petroglyphs has remained at a standstill. Various conservation steps were undertaken on the Chilpo-ri petroglyphs in 2008 and 2009, but these actions focused mostly on the removal of lichens. In addition, not only is there a lack of facilities such as information panels, protective measures and visitor pathways, a formal investigation has yet to be carried out on the exact number of rock art panels in the Chilpo-ri area. Therefore, it can be argued that the research, conservation management, and use of the Chilpo-ri petroglyphs must be preceded by an investigation into their exact number, location, and state of preservation. It is only on the results of such an investigation that a comprehensive conservation management program can be established and implemented.

A comprehensive conservation management program must balance the need for academic research, conservation management, and utilization of the petroglyphs. These three components are necessary requirements in any attempt to augment the value of our cultural heritage. It is important that the establishment of such a comprehensive program is carried out not solely by government administrators but through the formation of a committee (such as the "Committee for the Conservation of the Chilpo-ri Petroglyphs") that can promote the integrative participation of specialists from various fields. In addition, the examination of case studies from within and outside of Korea will allow a more efficient implementation of the conservation program. Efforts must also be made to include the local community (through active attempts at communication) when establishing such a comprehensive program – the importance of their participation must be acknowledged and attempts at conservation must take into account their opinions. Without the cooperation of the local community, the conservation of our cultural heritage is nigh impossible; upmost consideration should be given to the local community so that they may be the first to enjoy the fruits of the cultural and economic value of our cultural heritage. The time is now ripe for the conservation and management of the Chilpo-ri petroglyphs, an endeavor that must be accompanied by much consideration and discussion.

BIBLIOGRAPHY

Bak, Jeong-gun. 2002. "Amgakhwa yeongu reul wihan gogohakjeok jeopgeun bangbeob," *Amgakhwa ui josayeongu bangbeob gwa bojon bangbeob,* Proceedings of the 2002 Spring Conference of the Korean Petroglyph Research Association.

Cheongdo-gun. 2009. *Bullyeong-sa jeontab haeche bosu gongsa bogoseo.*

Fernandes, A. P. B. 2007. "The Conservation Programme of the Côa Valley Archaeological Park: Philosophy, Objectives and Action," *Conservation and Management of Archaeological Sites,* 9(2), pp. 71-96

Fernandes, A. P. B. and J. D. Rodrigues. 2008. "Stone consolidation experiments in rock art outcrops at the Côa Valley Archaeological Park, Portugal," Rodrigues J. D. and J. M. Mimoso eds. *Stone Consolidation in Cultural Heritage: Research and Practice; Proceedings of the International Symposium,* Lisbon, 6-7, May 2008, Lisbon: LNEC (Laboratório Nacional de Engenharia Civil), pp. 111-120.

Kim, Il-gwon. 2006. "Yeongil Chilpo jiyeok eui byeoljari amgakhwa yeongu," *Hanguk amgakhwa yeongu* 7/8.

Korean Petroglyphs Research Association and the City of Pohang. 2007 *Pohangsi Chilpo Sinheung-ri Amgakhwagun jipyo josa bogoseo.*

Yi, Jae-gyeong. 2006. "Yeongil-man ildae ui gogohakjeok seonggyeok," *Hanguk amgakhwa yeongu* 7/8, pp. 29-43.

03

A Case Study on the Conservation of the Petroglyphs of the Mont Bego Site, France

Focusing on the Stela "Le Chef de Tribu"

Young-hee Park

Location of the Site

Mont Bego, known for its Bronze Age petroglyphs, is located approximately 80 kilometers north of the seaside city of Nice, in the southeastern region of France (Fig. 1). It belongs to the commune of Tende in the department of the Alpes-Maritime. The petroglyphs are located at an altitude of between 2,000 to 3,100 meters.

Discovered in the 15th century, the petroglyphs of Mont Bego have been intensively investigated and studied. More than 100,000 rock engravings have been documented.

Among the many petroglyphs of Mont Bego, this paper focuses on an engraved rock called Le Chef de Tribu, its removal to the Museum of Merveilles and conservation. It also looks at the production of a copy that has been placed in the place where the Le Chef de Tribu originally stood.

Local Division of the Mong Bego Site

Most of the Mont Bego petroglyphs are on shale rock surfaces located at an altitude of 2,000 to 2,700 meters. The rock art is distributed over an area of

Fig. 1. Location of the Mont Bego site (Photo: Le Grandoise et le Sacré, p. 12)

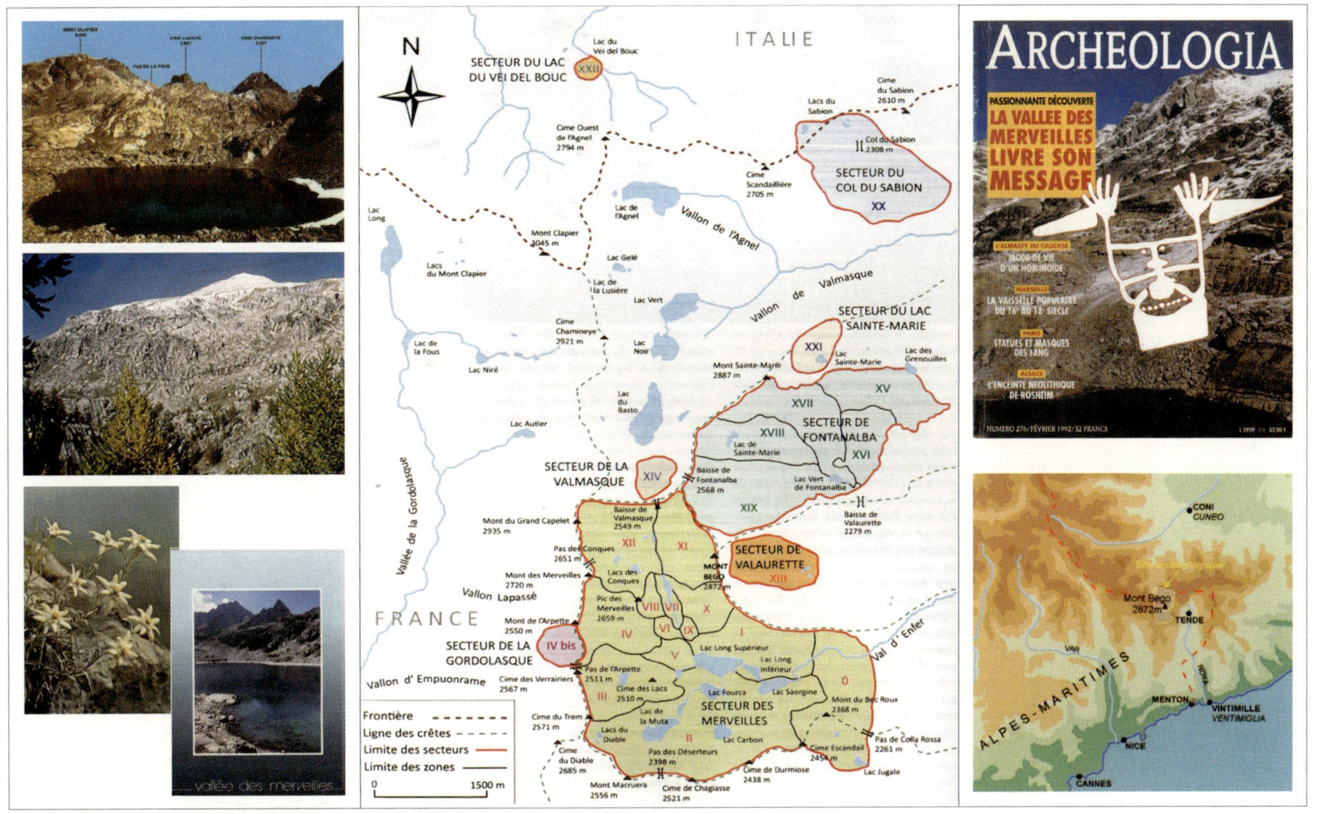

Fig. 2. Eight sectors of the Mont Bego site (Photo: La Montagne Sacrée du Bego, p. 13)

about 1,500 hectares that covers several mountain valleys. The smooth rock surfaces on which the engraved petroglyphs, both single figures and groups of figures, were produced were created through glacial erosion.

For analysis and research, the Mont Bego petroglyphs have been divided into eight sectors comprising 20 zones (Fig. 2) and each of the zones has been further divided into smaller sections. Individual names have been given to the rocks featuring petroglyphs.

Each petroglyph was investigated. As part of the recording process, each figure, along with the rock surface containing it, was given a number. A name was also given to the sector to which the rock art belonged. Based on the investigation results, the Mont Bego petroglyphs were divided into twelve categories according to theme.

Research History of the Mont Bego Petroglyphs

The petroglyphs of Mont Bego, which can be found throughout the Vallée des Merveilles (Valley of Marvels), were first brought to the public's attention by Pierre de Montfort in fifteenth-century, who travelled throughout the Tours region. Subsequently, the Mont Bego petroglyphs came to be investigated and studied by various individuals. The English botanist Clarence Bicknell carried out research on the petroglyphs of Fontanalba area of Mont Bego, and the Italian sculptor Carlo Conti produced a distribution map of the petroglyphs in the region.

Studies on the Mont Bego petroglyphs were based, above all, on primary research made on-site. As the most basic step of this on-site research, an exact copy of the rock art images was made (Fig. 3). The size and engraving method of each petroglyph figure was also recorded. A 1:10 ratio scale drawing of each image, detailed information about the rock on which the petroglyphs were engraved, and photographs of all the petroglyphs were also produced.

The on-site research was conducted in July and August and was accompanied by the examination of previously identified sites and surveys carried out in new areas by various institutes and museums.

Figs. 3. Producing copies of the Mont Bego petroglyphs (Photo: Le Grandoise et le Sacré, p. 24)

Classification of the Mont Bego petroglyphs

The petroglyphs of the Mont Bego site can be classified into the following categories according to theme: knives and daggers, halberds, weapons and tools, corniformes, ploughs and carts, T-signs, anthropomorphs, net-like geometrical figures, crosses and stars, and unclassifiable images (Fig. 4).

Of these rock art images, the most frequently observed motif among the Mont Bego petroglyphs is the "corniforme," which appears to be a head of a horned ox. Corniformes are depicted individually or in groups. Some appear alongside other motifs, thereby embodying new meaning. When a pair or two pairs of corniformes are depicted as being linked to one another, is interpreted as a representation of a plough or cart motif. This type of motif is also found in association with anthropomorphs among the petroglyphs of the Fontanalba Valley in the Mont Bego area.

Fig. 4. The petroglyphs of Mont Bego (Photo: Merveilles, un site, un patrimoine, un musée p. 5)

Interpretations of the Mont Bego Petroglyphs

Corniformes are the most frequently observed type of image among the Mont Bego petroglyphs and they have been interpreted as symbolic representations of wishes for fertility in a farming society. Such corniforme figures can be found engraved alongside net-like geometrical figures and anthropomorphs, as well as many dots. A good example of this is Rock 11 A of Group 2, Zone 4 which illustrates how corniforme figures developed into anthropomorphs (Fig. 5). Of the Mont Bego petroglyphs, the images that are regarded to be of the greatest importance, such as Le Sorcier (the Sorcerer, ZVIII GII R3[4]), La Déesse Acéphale (the Headless Goddess, ZIV GIII R16D), and the anthropomorphic figure with zigzag-shaped arms (ZIV GIII RI6D), are all based on the corniforme motif.

Fig. 5. The petroglyphs of Mont Bego (ZIV GII R11A) (Photo: Le mont Bego. Vallées des Mervielles de Fontanalba, p. 68)

Le Chef de Tribu

This rock, which contains 39 engravings, is located in the Vallée des Merveilles approximately 15 meters downstream from the Lac des Merveilles (Fig. 6). It is gray shale with bluish-green tones. The rock is 165 centimeters tall and it has a base width of 123 centimeters. Its greatest thickness is 31 centimeters and it weighs 1,070 kilograms. The front side of the rock is convex in form and its sides are rounded. It is believed that this rock came to be placed in its original position when the glaciers retreated during the Last Ice Age.

Due to the presence of several types of damage marks marring its surface, for conservation purposes a historic decision was finally made to move Le Chef de Tribu from its original location. The rock was moved on October 15, 1988 by a helicopter (Puma SA330) provided by the French Army. Afterward, an exact replica of Le Chef de Tribu, identical to the original in terms of size, weight, and acoustic conditions, was erected at its original location on October 24, 1988. The French Army's Puma SA 330 helicopter was also used to transport and place the replica. This process is illustrated, in part, in figures 11 to 25.

The Le Chef de Tribu is located at an altitude of 2,290 meters and several of the Mont Bego petroglyphs of greatest significance, such as Le Sorcier (2,350 m), La Déesse Acéphale' (2,470 m), the anthropomorphic figure with zigzag-shaped arms (2,470 m), are located in its vicinity, with the summit of Mont Bego acting as a focal point for these petroglyphs (Fig. 7). Thus, along with the Roche de l'Autel (Altar Stone, ZXI GI Autel), this area is of important meaning to the Mont Bego site.

Fig. 6. Le Chef de Tribu (ZVII GI R8)

Damaged Portion of the Le Chef de Tribu

Many damage marks can be observed on the surface of the rock Le Chef de Tribu. The left illustration of Fig. 9 shows damage marks made by an iron rod and the right illustration of Fig. 9 shows damage caused by the fracturing of the rock surface. Damage resulting from graffiti is also visible. The left illustration of Fig. 10 shows graffiti made in the Middle Ages and the right illustration of Fig. 10 shows graffiti made recently.

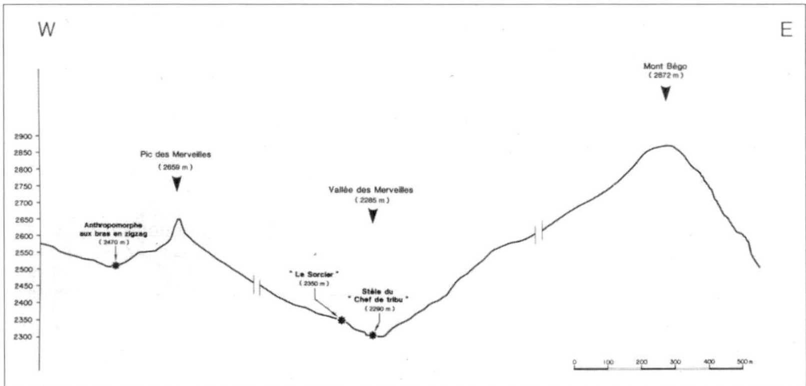

Fig. 7. Altitude of Le Chef de Tribu site (Photo: L'Anthropologie, Paris, Tome 94, p. 8)

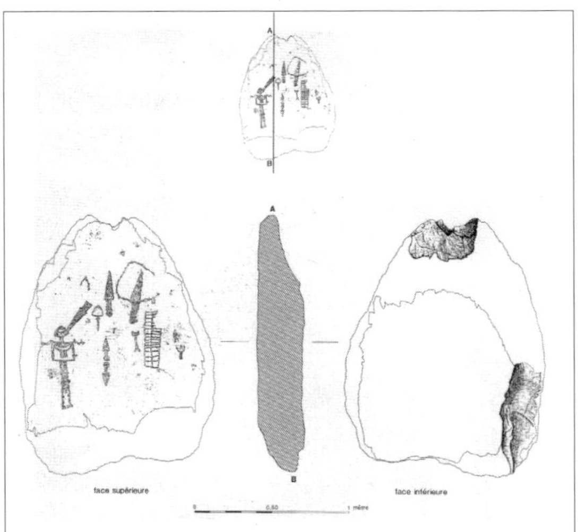

Fig. 8. Front and back side of Le Chef de Tribu (Photo: L'Anthropologie, Paris, Tome 94, p. 16)

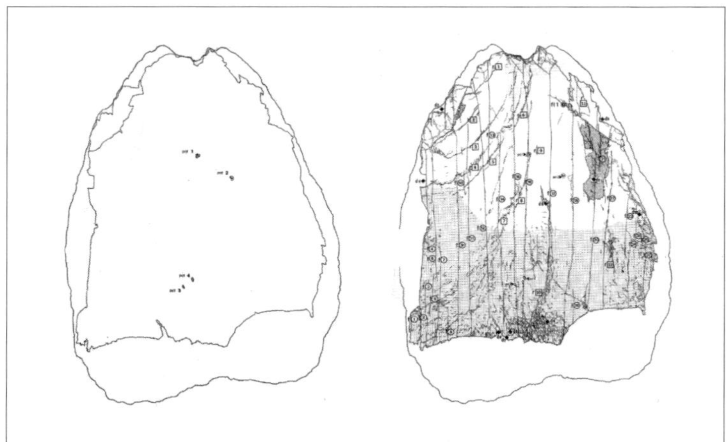

Fig. 9. Damage on Le Chef de Tribu (Photo: L'Anthropologie, Paris, Tome 94, pp. 18-19)

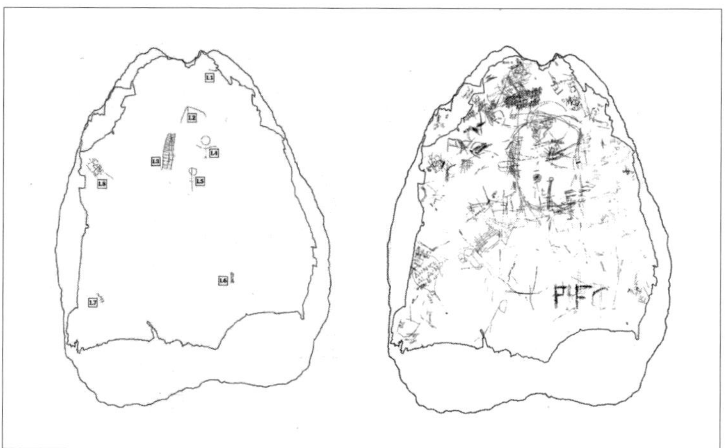

Fig. 10. Damaged marks found on the Le Chef de Tribu (Photo: L'Anthropologie, Paris, Tome 94, pp. 53-58)

Reproducing a Replica of the Le Chef de Tribu

Because of the continuing damage done to the rock as it stood in its original location, a replica of the rock was made, which was one of several actions that were carried out for the conservation of the Le Chef de Tribu. The first step of the reproduction process was to cover the original rock in latex, which was hardened in situ and then removed. This was used to make a cast for the replica of the rock (Fig. 11). The next step was to protect the original rock using various materials prior to transporting it to the museum (Figs. 12-15). A metal structure, nets and wooden planks used as handles were assembled around the rock to ensure its safe and scientific removal (Figs. 16, 20). The transportation of the Le Chef de Tribu to the Musée des Merveilles was perfectly executed and it has been exhibited there since 1996, the year of inauguration.

Figs. 11-12. Covering the surface of the Le Chef de Tribu with latex (left); Covering the front side of the Le Chef de Tribu with the first layer of latex (right) (Photo: L'Anthropologie, Paris, Tome 94, p. 61, 66)

Figs. 13-14. Finished rock cover consisting of latex and tarlatane muslin (left); Le Chef de Tribu completely muslin (right) (Photo: L'Anthropologie, Paris, Tome 94, p. 66)

Figs. 15-16. Attaching synthetic resin and felt around the rock (left); Wooden handles and polyester attached to completely covered rock (right) (Photo: L'Anthropologie, Paris, Tome 94, p. 66, 67)

Figs. 17-18. Moving Le Chef de Tribu in a metal structure (top); A net placed beneath covered the Le Chef de Tribu (bottom) (Photo: L'Anthropologie, Paris, Tome 94, p. 68)

Figs. 19-20. Establishment scene of metal structure and net (left); Le Chef de Tribu, covered completely, prior to its transportation by helicopter (right) (Photo: L'Anthropologie, Paris, Tome 94, p. 68, 69)

Figs. 21-22. Le Chef de Tribu being transported by helicopter, Super Puma SA 330 (top); Le Chef de Tribu being transported to the Musée des Merveilles (bottom)

Fig. 23. Le Chef de Tribu, October 15, 1988

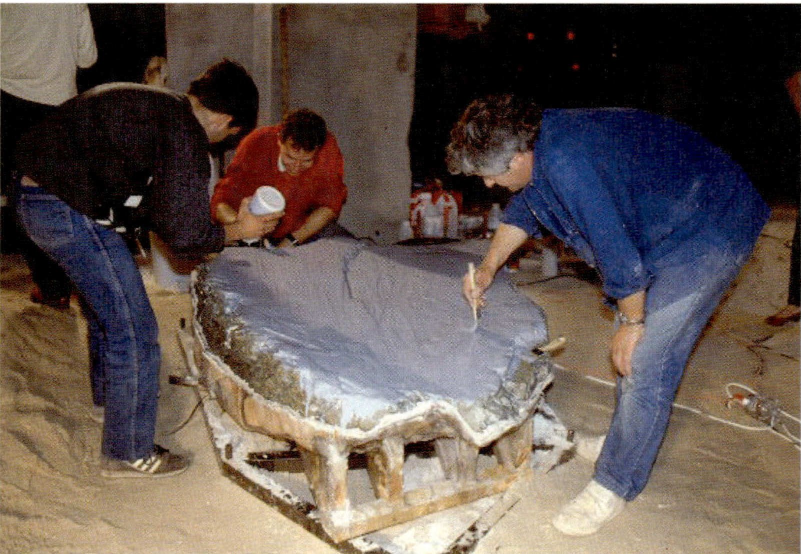

Fig. 24. Making a copy of the Le Chef de Tribu at the Musée des Merveilles

Fig. 25. Transporting the duplicate of the Le Chef de Tribu rock to the Vallée des Merveilles

BIBLIOGRAPHY

Maps

Carte Topographique (1:25 000) 3741 ouest st-martin-vésubie / col de turini (parties des feuilles au 1:50 000 de St-Martin-Vésubie et du Boréon) Institut Géographique National, France.

Carte Topographique (1:25 000) 3841 ouest breuil-sur-roya tende (parties des feuilles au 1:50 000 de Tende, de St-Martin-Vésubie, Du Boréon et de Viévola) Institut Géographique National, France.

Carte France géologique (1:2 500 000e) Pomerol Ch. et Petzold M, Masson, Paris, 1980.

Books and Articles

Conseil Général des Alpes-Maritimes. 1998. *L'échelle de Paradis;* Clarence Bicknell et la Vallée des Merveilles, p. 80, pl., photo., dessin.

Conseil Général des Alpes-Maritimes. 1998. *Merveilles, un site, un patrimoine, un musée,* p. 144, fig., photo.

Faton A. (Dir.). 1981. La Vallée des Merveilles, *Préhistoire et archéologie,* n. 27, fév., 1981, pp. 19-38.

Lumley H. de. 1982. *Origin et Evolution de l'Homme,* Laboratoire de Préhistoire du Musée de l'Homme, p. 256, pl., photo.

_________. 1984. Les gravures rupestres de l'âge du bronze de la Vallée des Merveilles, mont Bego, Alpes-Maritimes, *L'Anthropologie(Paris),* t. 88, n° 4, pp. 613-647, 27 fig., 1 tabl., 22 réf., bibl., résumés en français et anglais.

_________. 1991. Les gravures rupestres protohistoriques de la region du mont Bego, *Le mont Bego, Une montagne sacré de l'âge du Bronze. Sa place dans le context des religions protohistoriques du bassin méditerranéen.* Colloque International, Tende, Alpes-Maritimes, vendredi 5 au jeudi 11 juillet 1991, t. 1, pp. 82-94, 5 fig.

_________. 2013. Quelques réflexions sur la protection et la conservation des roches gravées préhistoriques et protohistoriques, *manuscrit,* p. 2.

Lumley H. de, avec la collaboration de J. Begin-Ducornet, A. Echassoux, N. Giusto-Magnardi, M.-A. Lumley de, P. Machu, Y.-H. Park, O. Romain, S. Saguez, T. Serres, et F. Villain-Rinieri. 1992. Le mont Bego. Vallées des Mervielles de Fontanalba. *Guides archéologiques de la France.* Ministère de la Culture.

Imprimerie nationale Editions, p. 165, 182 fig., dont 74 photo en couleurs et 4 cartes, 2 tabl., 65 réf. bibl.

Lumley H. de, avec la collaboration de J. Begin-Ducornet, A. Echassoux, A. Fournier, N. Giusto-Magnardi, G. Lavigne, M.-A. Lumley de, P. Machu, L. Mano, Y.-H. Park, O. Romain, S. Saguez, T. Serres, et F. Villain-Rinieri. 1995. *Le grandoise et le Sacré.* Gravures rupestres protohistoriques et historiques de la région du mont Bego, 1995, Editions Edisud, 452 p., 279 fig.(dont 14 pl. h. t.), 142 réf. bibl. Pochette de 14 planches dépliantes hors texte(fig. 266 à 279) et d'un livret:Itinéraire de découverte, précédes de quelques contes et légendes de la Haute-Roya, p. 32, 4 cartes, 3 tabl.

Lumley H. de, et A. Echassoux. 2011. *La Montagne Sacrée du Bego: Préoccupations économiques et mythes cosmogoniques des premiers peuples métallurgistes des Alpes méridionales,* CNRS Editions, p. 364, 377 fig., bibl.

Lumley H, de, et B. Pichard. 1983. Gravures de l'âge du bronze dans la Vallée des Merveilles. *In Catalogue de l'exposition "Art préhistorique des Alpes occidentals,"* Muséum d'Histoire Naturelle, Palais Longchamp, Marseille, mars-avril 1983, pp. 4-14, 31 fig.

Lumley H. de, J. Begin-Ducornet, A. Echassoux, N. Giusto-Magnardi, O. Romain, avec la collaboration de M. Dauvois, A. Fournier, G. Martini, et Noline. 1990. La stèle gravée dite du "Chef de Tribu" dans la région du mont Bego, Vallée des Merveilles, Tende, Alpes-Maritimes, *L'Anthropologie(Paris),* Tome 94, n° 1, pp. 3-62, 63 fig., bibl.

Lumley H. de, et L. Clergue. 2002. *Fascinant Mont Bego,* Editions Edisud, p. 142, pl., photo.

Lumley H. de, M.-E. Fonvielles, J. Abelanet. et al. 1976. *Livret-guide de l'excursion C1. Vallée des Merveilles.* Préface de José Balarello, p. 185. *IXe Congrès de l'Union Internationale des Sciences Préhistoriques et Protohistoriques,* Nice, 13-18 septembre 1976. Publié avec le concours du secrétariat d'État à la culture, Service des fouilles et antiquités.

Lumley H. de, R. David, F. Echassoux, C. Lanoux, et L. Mano. 1990. Enlèvement de la stèle gravée de l'âge du bronze ancien dite du "Chef de Tribu," *L'Anthropologie(Paris),* Tome 94, n° 1 pp. 63-84, 30 fig.

Masson E. 1992. La Vallée des Merveilles livre son message, *Archéologia,* n° 276, fév., 1992, pp. 12-23, photo., bibl.

Park Y. H. 1999. Une journée gracieuse au Musée des Merveilles, Tende 19 oct., 1999, *manuscrit et dessin,* p. 7.

Piggot S. 1983. *The earliest wheeled transport from the Atlantic coast to the Caspian Sea.* Thames and Hudson, London, p. 272.

Pomerol Ch. 1980. *France géologique, grands itinéraies avec une carte géologique au 1/2 500 000,* Editions Masson, 254 p., 100 fig., orientation bibliographique, index géologique.

04

Foz Côa:

The Story of Safeguarding a Prehistoric Site in Portugal

Mounir Bouchenaki

The Foz Côa site is situated in the valley of the Côa River, nearby the confluence of the Côa and Douro rivers, in the northeastern part of Portugal, not far from the Siega Verde in Spain. A famous open-air site dating from the Paleolithic period, it is inscribed on the UNESCO World Heritage List.

The story I want to present about the safeguarding of this site started a long time ago, in 1949, when Hidroelectrica do Côa announced a hydroelectric project to exploit the Côa River. This project, which included the construction of a dam, was revised successively in 1959, 1977, 1986, and 1989 and then finally adopted in 1991.

An impact study of the dam construction was commissioned in 1989 and the report issued thereafter by Professor Francisco Sande Lemos on the archaeological impact of the dam recommended that additional studies be undertaken. With the start of the dam project, the Côa Valley was surveyed and thousands of engraved drawings of horses and bovines as well as human and abstract figures were discovered.

EDP (Energias de Portugal, formerly Electricidade de Portugal), the firm in charge of building the dam, proclaimed in the presence of a public audience in November 1991 that "it has the duty to preserve the memory of archaeological remains." Therefore, in March 1993 it signed an agreement with the Portuguese Institute of Archaeology and Architecture (IPPAR) to conduct a detailed study of the area where the dam Vila Nova de Foz Côa was to be completed by 1998.

Portuguese archaeologists conducted a systematic survey of the area and discovered some amazing rock-engravings. Indeed, the survey of the lower part of the Foz Côa Valley revealed numerous groups of important Paleolithic, Neolithic, and Chalcolithic carvings. When local archaeologist Dr. Nelson Rabanda showed some of the petroglyphs to the press, a "major scandal ensued," as explained in an ICOMOS document published in 1995 by Professor Jean Clottes, then the chairperson of the ICOMOS Prehistoric Rock Art Committee: "N. Rebanda and IPPAR (The Portuguese Institute of Archaeology and Architecture) in charge of archaeological matters were accused of remissness by other archaeologists and by the press and a polemic developed about stopping the dam. There even was a heated debate in Parliament! This is probably the first time in Europe that the preservation of a rock art site has raised such interest and passion."

Despite the studies and discoveries made by the prehistory experts, construction of the dam started in October 1994, which led to a major protest by students and the general public. Due to the protest, the Portuguese authorities invited a UNESCO mission led by Professor Jean Clottes of the ICOMOS to visit the site in December 1994 to confirm the importance of the rock engravings. Clottes proclaimed that "the Côa Valley is the biggest open air site of Paleolithic art in Europe, if not in the world."

However, as Clottes also noted that "the rising water may protect the engravings from vandalism," construction of the dam continued amid continuing public protests. The Portuguese government came under pressure from the general public as well as from the international community. Indeed, the project and the dispute inside Portugal about the building of the dam and its catastrophic consequences for the rock art were mentioned during the months of November and December 1994 in major international newspapers such as *The Sunday Times, The New York Times, and The International Herald Tribune,* as well as by public broadcasters such as BBC.

On December 29, 1994, Ambassador of Portugal to UNESCO José Antonio Moya Ribera informed Federico Mayor, Director General of UNESCO, about the concerns caused by the dam project and requested the international organization to send a mission to Portugal in order to advise IPPAR on the protection and safeguarding of the rock engravings located in the Foz Côa Valley.

I was then asked, as Director of the UNESCO Division of Cultural Heritage, to conduct a multi-disciplinary study of the rock art with a team composed of Dr. Alain Bouineau, Director of the CEBTP of France (Centre of Study for Buildings and Public Works), Jacques Brunet, Conservator in the LRMH (National Research Laboratory on Historic Monuments) and Professor Philippe Malaurent, a hydro-geologist from Bordeaux University. We went to Lisbon on January 29, 1995 where we had meetings with Professor Nuno Santos Pinheiro, IPPAR President, and his colleagues. We then went to the site of the rock carvings in Foz Côa. Over the next five days we undertook technical visits as well as had meetings with several officials from EDP, the municipality of Vila Nova de Foz Côa, and archaeologists from IPPAR. At the end of our mission, the UNESCO team drafted a preliminary note summarizing the major issues that needed to be addressed and submitted it to IPPAR before leaving Foz Côa.

Then, and this was quite exceptional, the team was invited to spend another day in Portugal in order to have a meeting with the President of the Republic of Portugal Mario Suares at the Presidential Palace in Lisbon on February 4, 1995. President Suares told us that the tension in the country was very high about the dam project with demonstrations against it taking place every day in different cities across the country and, since our site visit was over, he wanted an opinion from UNESCO very quickly. He emphasized the importance of what was happening in a country "which is not so rich but where the population is willing to sacrifice economic benefits in order to keep the memory of its prehistoric times by not letting a dam flood important rock engravings." He showed us the slogan of a citizens group, "The carvings can't swim," which had been plastered on many secondary schools across the country.

As we were aware of the great importance of this issue which was creating a controversy for the President of Portugal and the Portuguese Government, we prepared the report in less than three weeks and sent it to the Portuguese authorities through the Permanent Delegation of Portugal to UNESCO.

In summary, the first proposal in the detailed report was to first create a site museum in order to have a research and information center on the site that would include a data bank on all the rock engravings, whatever the final decision—"dam or no dam." And it should be made in parallel with the creation of a national natural park that would preserve the beauty of the environment of the Côa Valley and be open to cultural tourism, thereby generating resources which could compensate for the cancellation of the dam construction.

The mission highlighted the uniqueness of the site and the exceptional value of the presence of an ensemble of rock engravings dating from the Paleolithic period. It pointed to the 1968 UNESCO Recommendation concerning "the preservation of cultural properties put in danger by public or private works," which calls for protecting them in situ, and emphasized that Professor Joao Zilhao had already noted that if the dam were to be built, the eventual flooding of the valley would submerge the rock art at depths that, in places, would be more than 100 meters.

During the summer of 1995, following general elections in Portugal, the newly-elected Prime Minister Antonio Guterres announced, when he submitted his administration's program to Parliament, "construction work on the Foz Côa dam has been suspended until the value of the archaeological heritage it would flood can be adequately determined." He also made clear that, if confirmation of its worldwide importance was obtained, as he hoped, construction of the dam would be abandoned altogether.

This decision was totally in line with UNESCO's conclusions following the January-February 1995 mission. This was considered as "a première" in the world of conservation of Cultural Heritage given that an important economic project had been abandoned in favor of a cultural heritage site.

As stated in November 1995 by Professor Joao Zilhao, "The national and international protest to stop the Côa dam and preserve its rock art has won a tremendous victory. I should like to remind you, however, that all these decisions carry the implication, for the Portuguese tax payer, of a loss of 150 million U.S. dollars already spent on the work so far carried out on the dam. These were not easy decisions, therefore, and I believe that the new Portuguese authorities are to be strongly commended for their vision..."

The following Resolutions 4/96 and 42/96 of the Official Journal of the Portuguese Republic were adopted: To halt the construction of the dam that threatened to submerge the archaeological sites; to provide a fund of US $150 million to support the economic development of the region; and, to install an archaeological park. Additionally, on May 27, 1997, the Portuguese government decided to classify the Côa Valley Archaeological Site as a national monument. As a follow up to this decision, the site of Foz Côa was proposed to be inscribed on the UNESCO World Heritage List.

Public attention culminated in June 1997 when a visit to Foz Côa was organized with the participation of King Juan Carlos of Spain, the President of the Republic of Portugal Jorge Sampaio and the UNESCO Director General Federico Mayor. "I fully back a bid by Foz Côa for World Heritage status," said Director General Mayor in front of the population of Foz Côa gathered for this exceptional occasion.

This happened very soon after the dam project was abandoned. In December 1998 the World Heritage Committee inscribed the site on the basis of criteria 1 and 3, stating that "the upper Paleolithic rock art of the Côa Valley is an outstanding example of the sudden flowering of creative genius at the dawn of human cultural development. The Côa Valley art throws light on the social, economic, and spiritual life of the early ancestors of humankind in a wholly exceptional manner."

After the great success UNESCO had in assisting and supervising the International Campaign for the Safeguarding of the Monuments of Abu Simbel and Philae in Egypt during the 1960s, Foz Côa is another exceptional case where Cultural Heritage and its value was finally recognized and saved.

05

Preservation and Collateral Damage

Knut Helskog

I wonder what the people who painted, carved, incised, or pecked figures on rock surfaces would have thought if they knew about the efforts being made to preserve some of their rock art. Would they have understood and been proud and happy for the respect accorded them or wonder why anybody would preserve the figures, build visitor centers, museums and walkways and engage in active conservation of the rock art itself as well as the surfaces and the areas where they were made? Curators and researchers aim to ensure the long-term survival of the figures, to have good documentation of the rock art for future generations of managers, and to promote research on understanding what they are about. Some rock art has lasted for thousands of years and will exist for a long time to come. But a lot has disappeared, and more undoubtedly will.

The preservation of sites includes the direct care of the figures themselves and their associated environments as well as measures to prevent the destruction of sites by natural and human forces (e.g. Bjelland and Helberg 2007). In this paper I will focus on the preservation of the environment because parts of the environment can be related to the meaning of the figures. By environment I mean both the surface on which the figures exist, the immediate physical surroundings and the larger geographic landscape. The figures might related to the topography, structures, and colors on the surface where the figures are located, to surface orientation and structures, vegetation in the immediate surroundings, or features in the general landscape. It might be a combination of interacting features that need to be identified, although to draw a line between what is possible and practical to preserve and what is not, is a challenge and often impossible. In this process the decisions taken to preserve the rock art itself might lead to a conscious or unconscious destruction of features linked to the meaning of the art, hence named collateral damage.

Preservation

Most often preservation focuses on the figures themselves because they, as the cultural product, might trigger some type of response in humans. The figures might be perceived as beautiful and mystic representations of a culture that no longer exists, displayed in a way very different from the way art is displayed today. Is it a dream about ancestors or past populations who inhabited the land, or is it just academic curiosity to understand the possible roles and meanings the figures might have had in various contexts such as representations of animals and people in mundane and sacred narratives, in rituals or festivities of past societies or one's own ancestors, and so forth? Among some, the representations and narratives connect to living identities considered important to continue and preserve. Alternatively, even though no direct continuity can be traced, it might be considered important to understand the cultural content and processes within a geographic region as a part of recognizing how people prior to the present interacted and coped with the environment. So the reasons for preserving the figures are multiple and contextual including the presence of social entities or individuals that can control the allocation and implementation of the necessary resources. On the other hand, some responses are to suppress and destroy the stories of others in a process of eliminating and dominating past and present identities. The aim might be, for example, to enhance and integrate one's own and new identity in the context of acquiring and exploiting or settling in a new territory. In essence, communities must accept or require that monuments such as rock art should be cared for. But then not all sites are equal; size and content differ and some are better received and graded higher than others by the public and by researchers, depending on personal or community perception and aim. In essence, preservation is as much for whom as for what and how. The issues and reasons are many and complicated.

Preventive Measures

Preventive measure ideally means strategies to avoid damaging the rock art and the surrounding environment. One way is to close public access to the rock art and protect it from natural forces. The first is clearly possible while to combat the forces of nature involves measures for temporary protection, such as covers of artificial or natural materials (e.g. peat or sand). Or, like was done for the Lasceaux Cave in France, one might resort to making an exact copy for public display and close the original site because of the possibility of serious damage long time visitation could cause (Bahn 2010).

Another way to prevent damage is sustainable presentation, which ideally means that the authenticity of the rock art and the natural environment is maintained when allowing public visitation. This is clearly impossible as any construction of signposts and walkways, for example, is intervention. Rather, it becomes a matter of controlling the number of visitors and installations so that they do not surpass what the monument can absorb without being damaged. For some sites this might mean limiting the number of visitors or like the case of Lasceaux, closing the monument for public access and constructing nearby an exact copy as possible. In essence, it is the responsibility of the managers to prevent sites from being destroyed by natural and cultural erosion and retain as much as possible of their original character in their local and larger regional setting. The first step in such a process is documentation of the status of the rock art and the rock surfaces by archaeologists, botanists, geologists, and curators working together. Management and action plans should be made and research directed toward conservation and presentation and visitation control as well as toward understanding the meaning and roles the rock art once had and have at the present (Hygen and Rogozhinskyi 2012). It is imperative that management understand what the carvings mean and communicate as this understanding is essential to choose what to document and how to manage. For example, how will understanding the relationship between the rock art and landscape influence the scope of the area around the rock art panels to be protected, and how we choose to protect and present the sites in situ. In this respect, the problem might be to decide how much of the area beyond the actual rock art itself should be included as part of the site. Therefore, among others, it is essential that managers be given the ability and right to conduct research as a part of their responsibilities.

Direct Intervention

Direct intervention includes any activity which touches the figures and the surface on which they are located as well as the adjacent and larger surrounding environment which are directly or indirectly connected with the rock art. This would include any type of on-site conservation (e.g. Reutova 2009), construction activity or vegetation management to protect the rock art. The site is the rock art and the surface on which it was made plus any part of the surrounding environment which might be directly related to the meaning of the rock art. It has become increasingly evident that features in the rock surface and the adjacent environment might be important factors both for the location and the meaning of the rock art (Taavitsianen 1981; Lahelma 2008). In a sense, the surface might be not only a "veil" between the world of the living where the figures are made and beings in other worlds of the understood cosmology (Lewis-Williams and Dawson 1990) but include natural features that are "living" parts of the communication to which the figures are central. I am not suggesting that this was and is always the case but as communication is the essence, this communication was and is addressed toward someone (Helskog 2010). This someone might have been human or non-human and the communication might have been strictly verbal, involved a type of performance and included a gift or a sacrifice. The location may have been chosen because of the need for a special place or it was a special time requiring a special place where figures or scenes could be depicted.

General Environment

Rock art is bound to special places and it is to be expected that the reasons for choosing a location might have changed through time. Such places of communication are well known from the ethno-historic record and include large natural phenomena such as mountains, valleys or canyons or features such as specially shaped or located boulders, rocks, and trees, and culturally constructed features such as idols, tents, and various types of buildings. Therefore, the possibility that the communication of which the rock art was and is a part might have some connection to the rock surface and the general location as well as the broader environment (Deacon 1988, 2012; Tacon 2002) should be considered in deciding what and how to preserve the rock art. To assess the need and implement the right type of preservation, understanding this connection is imperative. For example, if the meaning of a rock art site was related to a holy place such as a mountain or sacred valley, should all construction activities in this area be stopped to preserve the original meaning? Or, if communication and the meaning appear to connect figures and structures in the rock surface, would it be possible or desirable to preserve the connection, even to stop the destruction caused by natural and cultural forces? The rock art confined to the waterfall at Nämforsen in northern Sweden and their location could have had some connection to the sound of the waterfall (Goldhahn 2002) and the shore area of the Bay of Bothnia. As the land rose and displaced the shore, the making of rock art ceased in the Bronze Age, approximately early in the first millennium BC (Gjerde 2010:357). However, whether the termination was connected with a general change in beliefs, practices or the physical and mythological change caused by the shore displacement is unclear. Another example is Sarmishsay, a small canyon in Uzbekistan with unusually pronounced geological folding where rock art was made on rock faces on both sides of the small river (Fig. 1) (Khujanazarov 2001). In Finland, for example, it has been noticed that in several cases natural rock formations resembling human-like faces have come to be considered as being connected with rock art (Taavitsainen 1981; Kivikas 1995; Lahelma 2008). The rock art found in Finland so far consists solely of paintings located on vertical surfaces above lakes, rivers, and other waterways. A direct connection to water is common in the rock art of Fennoscandia (Ling 2008; Helskog 2013) as well as in many other places in the world. In some regions a direct association is much less common, such as in Drakensberg, South Africa or Brandberg, Namibia, although rain making seems to have been one of the aims of the ritual communication (Blundell 2004). The reason for the association to different features in the rock surface undoubtedly varies from purely practical to the most intricate association with belief and communication, but needs to be considered when managing rock art. In some cases, the location and features could have been so important for attributing meaning to the figures that without this context the figures were useless for the communication. Understanding, preserving, and promoting such possible relationships are imperative when planning what specific rock art panels to give access to and how. These features, structures, and relationships need to be protected and maintained as much as the figures themselves.

Fig. 1. The figures are carved on smooth black rock surfaces on both sides of the small river in the canyon Sarmishsay in Uzbekistan. The sides of the canyon consist of strongly folded sedimentary rocks, and in some places the rocks stand out like undulating veins. There is a distinct concentration of carvings in the narrowest part of the canyon where there is little space for housing settlements. Undoubtedly, there is a relationship between the figures, the structures of the rock surfaces, the river and the canyon that embodies many meanings which are as important as the rock surface.

Sites

One would expect that the different dimensions demand different solutions and that resources will be weighed against other plans, undertakings, and a Cultural Heritage act. One example is the choice made not to construct special installations to preserve the petroglyphs at Bangudae in South Korea from seasonal inundation caused by hydroelectric regulation of the adjacent river. Instead, the public is led to a viewing point to observe the petroglyphs through binoculars from a distance while exact copies are professionally exhibited in an adjacent museum built solely for that purpose. Another example is from Alta in the north of Norway where the main area of the Hjemmeluft/Jiebmaluok'ta rock art and adjacent prehistoric settlements were purchased, walkways constructed, a museum built and staffed, and management and action plans were implemented to preserve and maintain site integrity and control sustainable visitation. Part of the reason for the solution in Alta was that given the connection between the rock art and the prehistoric settlements, it was decided that they should be treated as whole. This was quite a change from the original plan for the area which was to develop a new harbor and industrial and residential areas.

Good examples of where areas with rock art have been given special protection to preserve and present the art in its wider natural and cultural context can be found all around the world, in national parks or areas given protection due to the rock art itself. National parks such as Giants Castle in Drakensberg, South Africa, and Twyfelfontein or /Ui-//aes in Namibia have walkways and information centers to ensure that visitors and associated construction do not exceed what the environment and sites can tolerate in an effort to maintain as much authenticity as possible. Often the best known sites are those that have been awarded World Heritage Status or falls within the protection of an active lawful organized heritage management system, and those where the local population has a direct vested interest. In addition there are thousands of sites, neglected or in danger of destruction by activities that alter the natural environment. Those most in danger are normally those that are in or adjacent to inhabited areas while those that are away from centers of habitation are mostly left to themselves and change as the forces of nature take hold.

The point is that in the cases where it can be demonstrated that parts of the natural environment can be or are the reason for the location of rock art panels and the communication they were or are a part of (e.g. Lødøen 2013), one should discuss to what degree and how it is possible to preserve that relationship. Likewise, because of the possibility that there might be such features within the immediate environment of rock art sites the problem is where to position walkways and the associated information systems so that they do not damage or destroy the relationship.

Figures

Even though there is a need to look beyond the rock art itself when considering what and how to preserve, there is no doubt that it is the figures that are given most attention by researchers as well as the general public. Figures can be perceived from being breathtakingly beautiful to coarse and ugly, enchanting and captivating or hypnotizing, spellbinding, or scary, even uninteresting to some. On a global scale the figure-variation is immense and mirror parts of the world as people understood it. The fauna depicted is always selective and culture specific and includes examples of the local or regional fauna, human and non-human forms, cultural items, and natural phenomena and such. Within the different panel types, the form and attributes of the figures might be repeated, yet no two panels are identical in content or in the surface on which the figures are made. From a global perspective, I do not know if two identical panels of rock art and rock surfaces have ever been found and recorded.

One might wonder if the integration of the surface and the figures had or has a special meaning or if the makers simply chose a surface because it was convenient. In some cases it is quite evident that colors, cracks, and fissures were incorporated into the figures, such as can be seen in some panels at the World Heritage site in Alta, Arctic Norway (Fig. 2) (Helskog 2010, 2012), at Zalavruga in NV Russia (Gjerde 2010) or at Sallys Rockshelter in the Mohave Desert, United States where quarts cobbles were left as offerings in cracks adjacent to the figures (Whitley 2001). The figures appear to have been positioned with deliberate care in relation to the features in the rock surface. In such cases it might be a serious mistake to engage in any conservation activity which might change this connection.

In essence, rock surfaces might not be simply a surface on which to make figures or a veil between worlds but an active ingredient (organism), but was it always so?

Preservation and Collateral Damage

Preserving rock art involves numerous decisions which might physically

Fig. 2. In this composition a number of the figures are integrated into a dark structure shaped like the front of a bear. Inside the head of the bear there is carved a den, a bear being speared, a bear cub, a dog and a human-like figure. Outside to the left there are two bears watching what is happening. Sets of bear tracks connect the bears, the den, and another den to the far left with the rock surface above and to the sea, which would have washed onto the rock surface at the time the figures were made.

affect not only the figures themselves, but also features in the rock surfaces, the immediate adjacent and wider surrounding environment, all which might have been significant for the communication or stories connected with the rock art. In some cases the connection seems almost self-evident but I suspect that the number of missed associations is infinite because of a lack of awareness and research, and the lack of tools to discover them. On the level of the individual panel, topographic features, cracks, fissures as well as patterns of colors and striations, might be connected with the stories and meaning once attributed the figures. This again means that when building walkways, placing information signs or engaging in any conservation which might touch the surface and immediate environment and surroundings of the panel, there is a danger of unwittingly destroying features which were connected with the communication and the meaning associated with the rock art. Natural features in the general area of sites similarly connected might likewise be damaged or destroyed by modern construction activities. So, preservation is partly a question of awareness and partly a matter of different scales of involvement for the preservation of rock art sites and the associated communication, from preserving single sites to creating national parks. Choices are not only made on what is practical, political, and economically possible to accomplish but also the priorities of what should be presented and what will be destroyed as a result–collateral damage.

BIBLIOGRAPHY

Bahn, P. 2010. "Lascaux in Crisis. The Clock is Ticking," *Current World Archaeology Magazine,* April/May 2010, pp. 28-36.

Bjelland, T. and B.H. Helberg, eds. 2007. *Rock Art. A Guide to the Documentation, Management, Presentation and Monitoring of Norwegian Rock Art,* Oslo: Directorate for Cultural Heritage, Norway.

Blundell, G. 2004. *Ngabayo's Bomansland: San Rock Art and the Somatic Past,* Uppsala: Uppsala University Press.

Deacon, J. 1988. "The power of a Place in Understanding Southern San Rock Engravings," *World Archaeology* 20, pp. 129-140.

_________. 2012. "Expressing Intangibles: A Recording Experience with Xam Rock Engravings," B. Smith, K. Helskog and D. Morris, eds. *Working with Rock Art. Recording, Presenting and Understanding Rock Art Using Indigenous Knowledge,* Johannesburg: Wits University Press, pp. 15-23.

Gjerde, J. M. 2010. *Rock Art and Landscapes: Studies of Rock Art from Northern Fennoscandia,* Ph.D. dissertation, Tromsø: UiT the Arctic University of Norway.

Goldhahn, J. 2002. "Roaring Rocks: An Audio-visual Perspective on Hunter-gather Engravings in Northern Sweden and Scandinavia," *Norwegian Archaeological Review* 35(1), pp. 29-61.

Helskog, K. 2010. "From the Tyranny of the Figures to the Interrelationship between Myths, Rock Art and their Surfaces," G. Blundell, C. Chippindale and B. Smith, eds. *Seeing and Knowing: Understanding Rock Art with and without Ethnography.* Johannesburg: Wits University Press, pp. 168-187.

_________. 2012. "Samtaler med Maktene. En Historie om Verdensarven I Alta," *Tromsø Museum Skrifter,* XXXIII. Tromsø: Tromsø Museum.

Hygen, A. S. and Rogoshinskiy, A. E. 2012. "Rock Art Management: Juggling with Paradoxes and Compromises and How to Live with Them," B. Smith, K. Helskog and D. Morris, eds. *Working with Rock Art. Recording, Presenting and Understanding Rock Art Using Indigenous Knowledge,* Johannesburg: Wits University Press, pp. 3-112.

Khujanazarov, Muhiddin. 2001. "Petroglyphs of Uzbekistan," K. Tashbayeva, M. Khujanazarov, V. Ranov, and Z. Samashev. *Petroglyphs of Central Asia*

Samarkand: International Institute for Central Asian Studies, pp. 80-121.

Kivikas, P. 1995. *Kalliomaalaukset: Muinainen Kuva-arkisto,* Jyväskylä: Atena.

Lahelma, A. 2008. "A Touch of Red: Archaeological and Ethnographical Approaches to Interpreting Finnish Rock Paintings," *Iskos 15,* Helsinki: Finnish Antiquarian Society.

Lewis-Williams, D. L. and Dawson, T. A. 1990. "Through the Veil: San Rock Paintings and the Rock Face," *South African Archaeological Bulletin,* 45, pp. 5-16.

Ling, J. 2008. "Elevated Rock Art. Towards a Maritime Understanding of Rock Art in Northern Bohuslän," GOTARC Series B, *Gothenburg Archaeological Thesis* 49, Gothenburg.

Lødøen, T. 2012. "Prehistoric Explorations in Rock: Investigations in and Beneath Engraved Surfaces," B. Smith, K. Helskog and D. Morris, eds, *Working with Rock Art: Recording, Presenting and Understanding Rock Art Using Indigenous Knowledge,* Johannesburg: Wits University Press, pp. 99-110.

Reutova, Marina A. 2009. *Petroglyphs in Sarmishsay. The Methodical Recommendations for Conservation,* Samarkand: Institute of Archaeology, Uzbekistan Academy of Sciences.

Taavitsainen, J. P. 1981. "Löppösenluola Hällmålning i Valkeala," *Finsk Museum* 1979, pp. 11-16.

Tacon, P. S. C. 2002. "Rock-Art and Landscapes," B. David and M. Wilson, eds. *Inscribed Landscapes. Marking and Making Place,* Honolulu: University of Hawai'i Press, pp. 122-136.

Whitley, D. 2001. "Science and the Sacred: Interpretive Theory in the U.S. Rock Art Research," K. Helskog, *Theoretical Perspectives in Rock Art research, Instituttet for sammenlignende kulturforskning, Serie B. Skrifter,* Vol. CVI. Oslo: Novus forlag, pp. 123-151.

06

Conservation or Destruction:

Attitudes toward Rock Art Sites (The Case of Russia)

Ekaterina Devlet

Thousands of rock art sites have been discovered and investigated within the vast territory of Russia. And, many reports have been made about the particular appearance and historical and cultural characteristics of the prehistoric rock art in many of its regions (Devlet, E. and Devlet, M. 2005) (Fig. 1).

Fig. 1. Rock art areas in Russia

Nonetheless, the enormous territory of Russia, with its diverse landscapes and historical and cultural heritage, remains a blank spot in terms of UNESCO's World Heritage List. Aside from the Bashkir Ural natural and cultural complex, which includes the Shulgan-Tash (Kapova) cave well-known for its numerous Palaeolithic paintings, Sikachi-Alyan is the only open-air prehistoric rock art site among the many petroglyph sites in the Russian Federation even included in UNESCO's Tentative List of locations for inclusion on the World Heritage List. The ancient enigmatic rock art from Sikachi-Alyan is a unique scientific, historical, and cultural property not only for Russia but also for the whole world (Fig. 2). Rock art preservation now encompasses a wide range of multi-disciplinary issues including direct and indirect conservation measures, protective legislation, public awareness, preservation and management strategies, development and tourism, air pollution and environmental changes.

Flooded Sites

Aside from the numerous natural processes involved in rock art deterioration, the most disastrous damage has been caused by huge industrial projects. Rock art sites of paramount importance have been flooded and lost through the construction of hydro-electric power-stations on the major rivers of Siberia. Such excellent sites as Oglakhti and Mugur-Sargol on the Yenisei River, rock paintings and engravings on the Angara, a tributary of the Yenisei, and many others have been partly or completely submerged and even the reduction of water levels did not bring back the rock art. The current situation of rock art protection in Russia can be ascertained by reviewing accounts of a few of the many hydro-electric projects (Helskog, Lobanova and Hygen 2008; Devlet, E. 2008, 2010, 2012; Devlet, E. and Devlet, M. 2002; Kochanovich 2005; Mikashevich 2003, 2008, 2011a, 2011b; Perjakova and Sklyarevski 1997).

The preservation of rock art sites under threat of submersion has been a growing concern since the 1990s. No technical solutions have been devised for the removal and re-location of rock art and only a few rock art panels have been relocated and thus survive. The recording of rock art images [through drawings, photographs, or other means] has long been the only way to retain

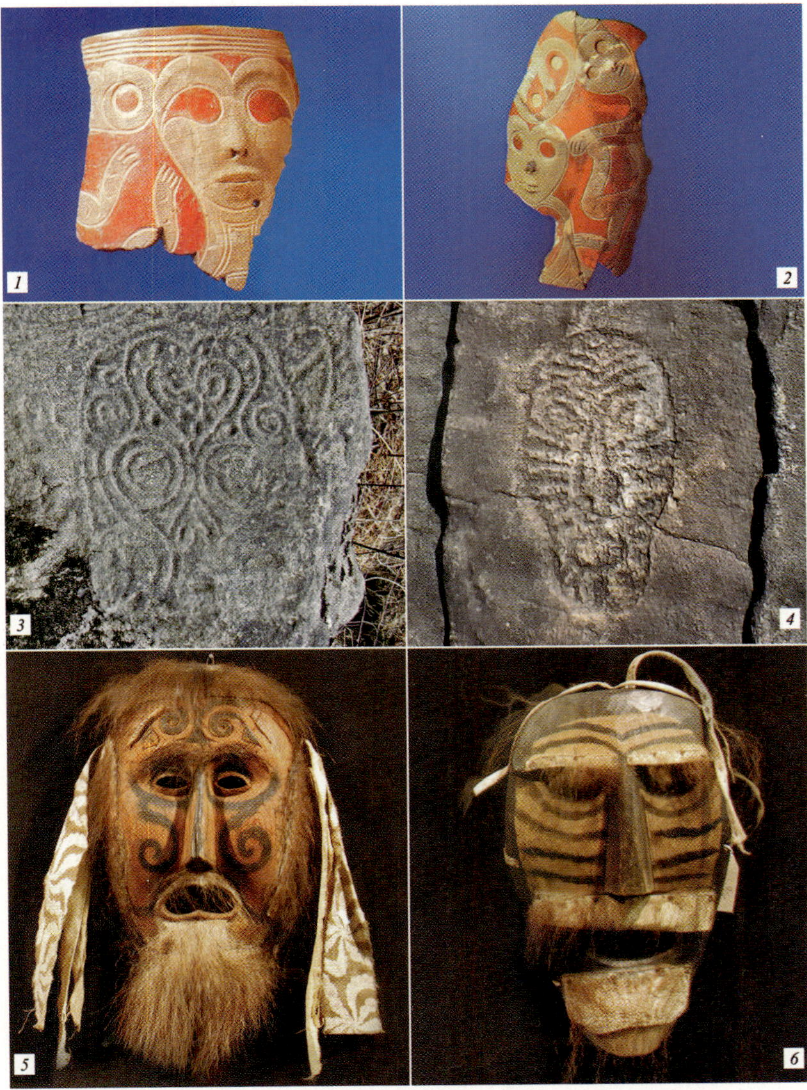

Fig. 2. Rock art in Sikachi-Alyan, Far East, Russia (UNESCO World Heritage Tentative List) in comparison with archaeological (1-2) and ethnological materials (5-6)

1-2 – Neolithic Ceramics, Voznesenskoye (Photo: Okladnikov, 1981, pp.18-19)

3-4 – Rock art in Sikachi-Alyan (Photo: Heritage Center in Khabarovsk)

5-6 – Masks of Nanai people from the Kahabarovsk Museum (Photo: Ekaterina Devlet) information for scientific studies. The state of the rock art sites flooded by the reservoir resulting from the construction of the Sayan hydro-electric power-station on the Upper Yenisei in Tuva and the state of the sites in the Middle Yenisei Basin that were submerged by the Krasnoyarsk reservoir were assessed differently. Following continuous observations, different results were reported in regard to rock art preservation after long-term exposure to water. The first example is that of the Mugur-Sargol and Aldy-Mozaga rock art sites in Tuva, where numerous images on boulders and cliffs on the banks of the Yenisei River in the Sayan Canyon and at other places have been destroyed. Outstanding images on Mugur-Sargol boulders were first partly submerged and then completely submerged by the reservoir created by the construction of the Sayan hydro-electric power-station (Fig. 3). After the submersion it was possible to observe this location during short periods when the water level was lowered but not one image could be found and intensive studies have only recently provided information on them (Devlet, M. 1980, 1998). When the water level was lowered it was not possible to see the images as all the pecked and engraved lines of the petroglyphs were filled with sediment deposited by the river. A similar situation involving the obliteration of rock art with seasonal river deposits can be seen currently at the Sikachi-Alyan rock art site and I shall discuss the situation further (Fig. 4).

Fig. 3. The Mugur-Sargol rock art site in the Upper Yenisei area, completely flooded in the 1980s by the reservoir of a hydro-electric power-station (Photo: Marianna Devlet) (left)

Fig. 4. Petroglyphs at the Sikachi-Alyan rock art site are subject to seasonal damage from water and ice action. (Photo: Ekaterina Devlet) (right)

Different observations were made in the Middle Yenisei, where rock art sites of prime importance such as Oglakhty, Sukhanikha, Tepsei, Ust'-Tuba, and Turan (Fig. 5) have been partly under water since the mid-1970s (Francfort and Sher 1995; Sher 1999; Sher et al. 1994; Miklashevich 2003, 2008). In the 1990s the water level in the Krasnoyarsk Reservoir began to drop and the Yenisei River returned to its original level after a few years. According to E. Miklashevich, a comparison of prior documentation and the state of the panels in the 1990s revealed that there was not much deterioration and almost all the recorded rock panels with petroglyphs and paintings survived being submerged. Unfortunately, the conclusion that submersion posed only a mild threat was short-sighted. Reexamination of the sites since the late 1990s has revealed additional deterioration. "The water eroded and destroyed the rock bases and washed away the soil; some low panels are crumbling to pieces, and though the upper panels are still preserved, they have become almost inaccessible to investigators because of damage to the rocks below and it is very possible that in the future they may be lost too." (Miklashevich 2003:101).

The problem posed by submerging rock art is well-known in Korea and observations made at the Bangudae rock art site are valuable for understanding the action of water in connection with the preservation of petroglyphs.

Fig. 5. A copy of the Oglakhty rock art panel with the most archaic rock art images from the Middle Yenisei region. The site was flooded in the 1960-70s by the reservoir of a hydro-electric power-station and destroyed after a drop in water levels (Rubbing by Elena Miklashevich).

Sikachi-Alyan

Sikachi-Alyan is located on the right bank of the lower Amur River in the area of an ancient Nanai village of the same name, which in the Nanai language Sikachi-Alyan means "Boar's Hill" (Figs. 6, 7). This is an area of lowlands, which has been protected from glacial action by the Sikhote-Alin ridge. In Russia there are several different rock art areas with special stylistic features as well as historic and natural backgrounds. Sikachi-Alyan belongs to the rock art tradition of the lower Amur and Ussuri river basins (Devlet, E. and Devlet, M. 2005) and it is remarkable for its entirely original art, an independent center of prehistoric creativity with long-standing diachronic concepts incorporated in rock art motifs and preferences in style. Important new information on local rock art was obtained during rock art protection projects here.

The petroglyphs are mainly concentrated alongside the Amur River for a distance of 6 kilometers. They are on basalt boulders lying at the foot of a high terrace, on the sandy riverbanks, and even in the water.

The Sikachi-Alyan rock art is characterized by anthropomorphic mask-images of various shapes. Many of them are elaborate, stylized, and ornamental. They are diverse in form, size, and detail; oval, heart-shaped, trapezium-shaped and combinations of several of these forms, some with strikingly pronounced contours and some without. The motifs include animals such as elk, horses, tigers, and boars, anthropomorphic images, birds, snakes, cup-marks, and concentric circles.

Fig. 6. A rock art motif from Sikachi-Alyan. The boulder was re-located by ice. (Photo: Ekaterina Devlet) (left)
Fig. 7. River deposits on the decorated rock surfaces in Sikachi-Alyan (Photo: Heritage Centre in Khabarovsk) (right)

In recent decades rock art projects in the area have been focused mainly on documentation and protection of petroglyphs, which are endangered by the actions of the river, especially by ice drifts. A specific threat in Sikachi-Alyan is the constant moving of some of the boulders bearing petroglyphs as the river is a "floating," living system. At Sikachi-Alyan it was calculated that some of the boulders bearing petroglyphs can be moved by ice over a distance of up to 55 meters. This natural action may result in the appearance as well as the disappearance of some images. During regular surveys, new rock art surfaces, initially covered with river deposits, have been revealed along the river bank (Figs. 8a, b).

A new concept of rock art management and public access is currently under consideration. Arrangements already introduced are focused on rock art protection and new cultural practices in rock art display to increase public recognition and awareness of its inherent value.

An evaluation of the state of preservation of rock art sites has demonstrated that among the various natural elements of destruction, seasonal ice drifts, seasonal fluctuations in water levels, temperature changes, freezing, wind erosion, and vegetation are most at play here. As it flows, the river moves massive amounts of sand and silt, which, in turn, dislodge stones located in the flood zone. The images on the stones, which were temporarily covered by water, are barely visible after its retreat because of the silt which, under the influence of wind, rain, and sun, turns into a hard crust. The alluvial soil encourages the growth of plants and their roots expand the cracks in the

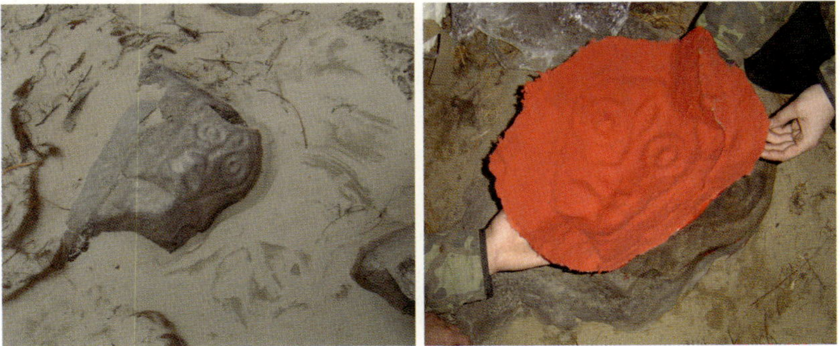

Figs. 8a, b. A bas-relief mask excavated from the sand (a, left) and a silicone matrix of the rock art (b, right) in Sikachi-Alyan (Photo: Ekaterina Devlet, Albert Babaev)

stone. The dark surfaces of the stones help to increase the temperature change between the shady and sunlit sides of the boulders by up to 30 degrees in the period of high solar activity (from April to August), which creates a 24-hour cyclical tension in the stone, and promotes the appearance of cracks (Laskin 2009).

The peculiarity of the Sikachi-Alyan rock-art site is the continual displacement of the boulders by the Amur River, with maximum movement during the spring ice drift. Because of the great depth and the speed of the current, ice plates up to 1.5 meters thick bump up against the basalt blocks and rocky ledges of the cape. At the same time, many stones overturn easily and are chipped from bumping against each other as they shift higher up or along the line of the drifting ice. As a result of this irreversible and recurring natural phenomenon, new images are revealed, while some that were documented earlier disappear.

Field investigations from 2000 through 2003 focused on the development of rock art protection projects, and a comparison of the current situation was made with drawings from the 1950s that were done by Okladnikov. Over a period of just 50 years, more than 25 stones with petroglyphs were overturned or moved over distances of between 0.2 and 55 meters. New rock art was discovered, while some of the previously documented images could not be seen. In 2000 it became obvious that at one of the locations about 10 boulders with petroglyphs had shifted from 2 to 10 meters over the last 50 years due to drifting ice; they had been moved or overturned, and several new carvings were discovered. In June 2003, the level of the Amur River was very low (-75 cm), which made it possible to clarify the location of the petroglyphs. It was found that large stones were mostly untouched, but smaller ones, located in places with a shallower water level, had shifted 3 to 20 meters, and some even as much as 55 meters and many of them are now in an inverted or modified position (Laskin 2009). Some boulders are now in changed positions, impeding access to this cultural heritage site.

The fact that boulders with images documented by Okladnikov (1981) could not be found, however, does not mean that they have been lost: they may have been simply overturned. While documenting the site, Okladnikov had drawn diagrams of two surfaces with mask-images, based on information from local people who had been unable to reveal them to him because they were hidden by deposits or overturned. This is grounds for considering deliberate re-location of some objects in immediate danger from the water and ice activity to where they will be safe but no farther than necessary from their original site.

The rock art site is also impacted by such negative anthropogenic influence and destruction as graffiti, fire making, and trashing of the area—the result not only of unregulated tourism, but also of the activities of the local inhabitants. Fishermen tie the metal parts of their fishing nets to boulders and fasten their boats to them, which leaves deep scars on the stone. A distinctive feature of the Sikachi-Alian site is that it is located in a village of the indigenous Nanai people of the region, and this presents fairly unusual problems for Russia in terms of traditional ownership. To this day, motifs similar to the Sikachi-Alyan petroglyphs exist in the traditional rituals and ornamental art of the Nanaitsy, Ulchi, Nivkhi, and Udegei tribes (Fig. 9). Local people use the boulders for their everyday needs, putting their food and drink right on the stones. Activity on or near some boulders can visibly change their surface. That is why the area needs to be fenced off but the local inhabitants are against fencing off the area because it is part of their cultural heritage. For local people easy access to the convenient places on the riverbank and short-term economic and everyday practical requirements might seem more important than preservation of the site. That is the main reason for the action aimed at involving the local community in the practical management of the site. Prospects for guaranteed occupation and financial support promote positive attitudes and interest in the preservation of the site.

Thus, among the urgent preservation needs is the proper fencing off of the area, at a distance from the rock art and hidden in the bushes. Another pressing need is permanent access for control aimed at the direct as well as indirect prevention of visitors touching decorated surfaces (Laskin et al. 2005; Medvedev 2011).

In the last three decades many things have changed in the area. One of the most positive events worth mentioning is the construction of a cultural center with a museum (linked to the State Museum named after N. Grodekov in Khabarovsk). The establishment of a modern cultural center and museum in the village of Sikachi-Alyan (Fig. 10a) has created favorable conditions for visitors and is encouraging the local people to get involved in rock art protection. It has also contributed to raising public awareness and recognition of rock art as a unique cultural heritage.

Fig. 9. Paintings of rock art motifs on houses in Sikachi-Alyan (Photo: Ekaterina Devlet)

Figs. 10a, b. The modern museum center (a, left) and information steles (b, right) in Sikachi-Alyan (Photo: Ekaterina Devlet)

A non-commercial foundation, The Historic Heritage of the Amur Region, has focused on the dissemination of knowledge on rock art heritage and the creation of signage as well as casts and displaying copies of petroglyphs (as of 2004, more than 20 silicone matrices of images had been made). Now they are part of the display in the museum in Sikachi-Alyan and are taken as exhibits to other towns. Made using modern technology, these casts can provide a source of information on local rock art and the culture of the Russian Far East in areas difficult to access. Collaboration of the department responsible for cultural heritage protection and funds management resulted in the construction of interpretation centers and steles marking the borders of the site (Fig. 10b).

As for the current stage of the site's protection, discussions with the local authorities and scientific community resulted in enhanced understanding of the priority of conservation management, which needs to be implemented in this new period of rock art protection and public awareness starting in 2013. There are several good arguments in favor of a new concept for rock art management and provision for visitors. Recent discussions about preserving the Sikachi-Alyan rock art site included arrangements for visitors, appropriate design, and steps for sustainable management and maintenance, which follow to a large extent Scandinavian patterns and demands with regard to public awareness. To ensure that this project develops smoothly, priority should be given to the implementation of protection measures. These should be aimed at protecting the site and guaranteeing its integrity and authenticity, providing a suitable welcome for visitors and involving local inhabitants in the process of rock art protection and related cultural activities as well as the promotion of tourism. Ongoing studies of the site may yield new information, which may help to update the cultural context.

The motifs carved on basalt boulders reflect a complicated system of beliefs and ongoing ritual practices. Sikachi-Alyan and other similar rock art sites could be sanctuaries for people from very remote areas who for generations have been coming to these places on special occasions to perform spiritual rituals. More advanced studies could help assemble an important information base that may result in historical re-attribution of the site. Numerous facts on rock art techniques and enhanced understanding of prehistoric thoughts and beliefs have resulted from preservation programs at various rock art sites.

Tomskaya Pisanitsa

The first attempt at rock art conservation and the establishment of an open-air museum at a Southern Siberian rock art site was undertaken in the Kemerovo region. Tomskaya Pisanitsa, a site recognized from over 290 years ago, became the heart of an archaeological park in 1988 and a museum was opened near the site in 1995 (Figs. 11a, b). The park includes improved viewing areas for visitors near the rocks with engravings, a museum building, an archives, reconstructions of ancient cemeteries and dwellings, ethnographic re-creations and other attractions (Martynov and Martynova 1993; Soleilhavoup 1994). The Rock Art of Asia Museum displays copies and pictures of the most distinctive rock art compositions from Siberia, Middle Asia, Mongolia, Korea, China, India, Pakistan, and other places. An important part of the work of the open-air museum's staff is tied in with the dissemination of knowledge about rock art and as a result almost all local inhabitants are aware of rock art and its importance. The Archaeological Park has won a number of awards for its activities aimed at making the region's cultural heritage better known.

The rock art site Tomskaya Pisanitsa, whose name means rock art from the Tom River, is located on the banks of the Tom River (Okladnikov and Martynov 1972; Martynov 1988). It should be noted that the Russian word *pisanitsa*, which is sometimes wrongly translated as "drawing," is applied to both paintings and petroglyphs in the Siberian case. The pecked and engraved petroglyphs of Tomskaya Pisanitsa are on schist rock and they are badly weathered and many have been vandalized. The first attempt at rock art conservation in Siberia was

Figs. 11a, b. Arrangements for tourist access to the Tomskaya Pisanitsa rock art site and an attempt to prevent water-flow from above with a concrete barrier, Tom River region, Siberia (Photo: Elana Miklashevich)

undertaken here, but attempts to stop deterioration of the surface (crack filling, surface impregnation) were not very successful. Conservation projects from the 1980s and early 1990s had lamentable and sometimes irreversible results as they even contributed to further rock art deterioration. The problem was made even worse by poor documentation of the conservation work that was carried out, particularly the methods and techniques that were used. In recent years a systematic program aimed at reducing the deterioration of surfaces, improving their poor state of preservation and mitigating the consequences of vandalism, severe Siberian weather and inappropriate conservation activities has been introduced. The team from the State Institute of Restoration removed some modified, unattractive material from cracks and experimented with artificial varnish for making graffiti less conspicuous and with other conservation techniques aimed at achieving pleasing results (Figs. 12-14a, b) (Ageeva and Kochanovich 2011; Miklashevich 2011b).

Figs. 12a, b. Decay of a rock art panel and degradation of repairs (a, left); new conservation at the Tomskaya Pisanitsa rock art site (b, right) (Photo: Elena Miklashevich, Alexey Kochanovich)

Figs. 13a, b. Decay of the rock art panel and degradation of conservation (a, left); cleaning of the cracks on Tomskaya pisanitsa rock art site (b, right) (Photo: Elena Miklashevich, Alexey Kochanovich)

Figs. 14a, b. Abrasions of recently carved letters (a, top) and artificial varnishing of the abraded areas (b, bottom) at the Tomskaya Pisanitsa rock art site (Photo: Alexey Kochanovich)

Besovye Sledki and the Zalavruga Rock Art Site

Another preservation project with a long history involves petroglyphs in the North European part of Russia (Fig. 15), namely the famous sites on the Vyg River in the vicinity of the White Sea near the town of Belomorsk (Savvateyev 1990, 1994; Lobanova 2009). The rock art here, which has been dated to between 4500 and 3000 B.C., became known through research undertaken in the 1920s. A flat surface with images known as "Besovye Sledki," meaning "Devil's Footprints," was covered over by a concrete building because a dam threatened the physical survival of the site (Figs. 16-18). This involved a wide rock outcrop measuring 12.5 meters by 7.8 meters with engravings including a central phallic figure of a mystical creature surrounded by numerous images of footprints, birds, and solar symbols. Years later the construction of this building was criticized in the wake of comparisons with a protective construction at the Peterborough rock art site in Canada (Wainwright I. 1985; Bahn, Bednarik and Steinbring 1995, 1996). This condemnation overlooked the fact that at the time the building was constructed in Karelia, a structure of this kind was the only way to avoid the physical destruction of the surface and its 470 engravings. Before the construction of the dam on the Vyg River, the place where the Besovye Sledki rock art site is located was known as Shoirukshin Island, but the area is no longer an island as it is no longer surrounded by water and thus the dam has threatened even the physical survival of the rock art site. We need to recognize that in the 1970s such a structure was the only way to preserve a surface with numerous engravings and avoid unfair criticism. The current situation with this structure, however, should be viewed as a lapse in cultural-heritage management on the part of the local authorities. The building deteriorated and was strongly criticized, the core problem being that the decay of the concrete endangers both the rock art and visitors. The building was closed to tourists and the rock surface was covered by sawdust, textiles, and sandbags to allegedly protect the rock art. This inappropriate action resulted in algae growth and the coloring of the rock fragments by pigments that bled from the textiles (Figs. 19a, b).

Fig. 15. Rock art from Onego Lake, Northwest Russia (Photo: Igor Georgievsky)

Another more promising example of preservation and the regulation of tourism in connection with rock art can be found at the Zalavruga rock art site (Figs. 20-24). This major concentration of rock art in the Belomorsk area attracts many tourists who come here on their way to Solovki, a major place of pilgrimage in the Russian North. Activities focusing on preservation, sustainable management, the promotion of public awareness and the dissemination of knowledge have been implemented at the Zalavruga rock art site (Helskog, Lobanova and Hygen 2008). In their report, the researchers declared that the joint Norwegian-Karelian project "Preservation of Karelian Petroglyphs" was the first international project dealing with Karelian rock art. A substantial amount of data on the status of the rock art was collected, including assessments of the degree to which it has deteriorated as a result of natural and anthropogenic processes. Moreover, new methods of conservation

Fig. 16. Deterioration of the building constructed above the Besovye Sledki rock art site, 2006 (Photo: Ekaterina Devlet)

Fig. 17. Inside the building at the Besovye Sledki rock art site, 2011 (Photo: Igor Georgievsky)

Fig. 18. Exhibition on rock art from Karelia in Finland. Copy of the rock art motifs from the Besovye Sledki and Zalavruga rock art sites by Svetlana Georgievskaya, Northwest Russia (Photo: Igor Georgievsky)

Figs. 19a, b. Besovye Sledki rock art panel is covered by a wooden construction to protect the petroglyphs (a, left); the removal of sandbags and textiles used in hope of protecting the petroglyphs (b, right) (Photo: Ekaterina Devlet, Nadezhda Lobanova)

were used and new petroglyphs were discovered in the Lake Onega area as well as in the lower reaches of the Vyg River. The project made it possible to develop new approaches to the documentation of rock art in Karelia, to promote rock art protection, popularization and education based on Norwegian research experience and achievements and to strengthen research links and mutual interest in Karelian and Norwegian rock art (Helskog, Lobanova and Hygen 2008:94-95). As a result of this collaboration, many positive changes in rock art preservation and public access have been made (Figs. 21-23).

On the Kola Peninsula in Northwest Russia many new and important sites were researched and documented in the 1990s and 2000s, following their initial discovery in 1997 (Kolpakov and Shumkin 2012). These include about 1,200 petroglyphs dating from the 4th to 2nd century B.C. Among the carvings are depictions of boats and white whales. A positive development is that access to the rock art on these islands is being properly controlled and a new visitor center has been established (Figs. 25-26).

Fig. 20. Zalavruga rock art site in the vicinity of the White Sea, NW Russia (Photo: Igor Georgievsky)

Fig. 21. Volunteers remove rubbish and plants from cracks of the Zalavruga rock art site in the vicinity of the White Sea (Photo: Nadezhda Lobanova)

Fig. 22. A wooden walkway at the Zalavruga rock art site (Photo: Nadezhda Lobanova)

Fig. 23. The walkways at the Zalavruga rock art site are not straight. (Photo: Nadezhda Lobanova)

Fig. 24. Rubbing of a rock art motif at the Zalavruga rock art site (Photo: Ekaterina Devlet)

Fig. 25. Kanozero rock art site, Kola Peninsula, Northwest Russia (Photo: Evgenii Girya)

Figs. 26a, b. Kanozero rock art site, Kola Peninsula, Northwest Russia (Photo: Evgenii Girya)

Rock Art Sites in the Upper Reaches of the Lena River and at Lake Baikal

The most important of the conservation programs was the supervised and permanently supported Center for the Preservation of Historical-Cultural Heritage in Irkutsk. It was started at the Shishkino rock art site (Okladnikov 1959, 1977) on the Upper Lena in 1987 and was expanded to the shores of Lake Baikal in 1992 (Okladnikov 1974) (Figs. 27-33).

Fig. 27. Shishkino rock art site, Upper Lena river, Siberia (Photo: Ekaterina Devlet)

Figs. 28a, b. Vandalism at the Shishkino rock art site by modern re-tracing in white of ancient paintings made in red pigment and its decay (Photo: Ekaterina Devlet, Elena Miklashevich)

A group from the Irkutsk Center of Historical and Cultural Heritage has been conducting a long-term field project to record rock art sites along the basin of the Upper Lena. Previously known rock art sites were re-recorded and some new compositions were brought to light. Among the motifs are anthropomorphs and hunting compositions belonging to various periods from the Bronze Age to the Medieval period (Melnikova et al. 2011).

The inter-disciplinary collaboration of chemist-conservators, biologists, geologists, physicians, and engineers ushered in a new stage in rock art protection projects in Russia (Bednarik and Devlet E. 1992; Ageeva et al. 1993; Perjakova and Sklyarevski 1997; Miklashevich 2011a). The introduction of international expertise in the field of rock art preservation slowly, step by step, led to it being recognized as a multi-disciplinary endeavor involving questions of protective legislation, public awareness, preservation and management strategies, direct and indirect conservation measures and how to deal with pressure resulting from development and tourism, air pollution and environmental changes. This inter-disciplinary approach made it possible to evaluate conditions for rock art preservation as well as conservation methods and priorities and to form a new view of the problems involved in recording

Fig. 29. Vandalism at the Shishkino rock art site by modern recoloring and scratching of the surface. A wide range of techniques used by a team of conservators (Photo: Elena Miklashevich)

rock art (Figs. 27-29).

Rock art research in the vicinity of Lake Baikal was pioneered over a century ago but new sites are still being discovered (Podovkina and Goriunova 1995). A. Okladnikov divided rock art images in the Baikal area into different periods dating from the Neolithic to medieval times. Some of them are connected with forest hunters, others with nomads from the steppes (Okladnikov 1974).

The most serious problems exist at the Sagan-Zaba rock art site and it is impossible to imagine a very happy future for it. According to recent observations it may be assumed that the deterioration was caused mainly by changing water levels in Lake Baikal after the construction of a power-station on the Angara River as well as by uncontrolled access by visitors, which is quite obvious. Vertical marble outcrops (12-15 m high) with depictions are now only 2 to 3 meters from the water and precipitation falling onto the rock is quite intense, resulting in rapid surface destruction with intensive algae growth between the grains of the rock and under the laminar exfoliated areas on the marble rock (Fig. 30). The damage is irreversible and there are no appropriate solutions for the long-term preservation of the site.

Fig. 30. This rock art panel on a vertical marble surface (to the right) is near the water in Sagan-Zaba Bay, Lake Baikal, Siberia. (Photo: Elena Miklashevich)

Rock engravings in Sagan-Zaba Bay on the shore of Lake Baikal were described and recorded for the first time in 1881 by N. Agapitov, in 1913 by T. Savenkov, and in 1968 and 1971 by an expedition led by A. Okladnikov (Okladnikov 1974). Comparing the new recordings of the rock art with those made more than a hundred years earlier, it is obvious that only a few figures have been partly or completely lost (Figs. 31-32). This shows that the collapse of the rock art panels was a long-term process and changes in the water level accelerated the process.

Images are now visible only on parts of the rock surface, and according to previous documentation other depictions were lost more than 115 years ago. So changes in water levels in the lake are not the main factor causing decay. It is assumed that the main factor was the loss of a protective "roof." It is easy to see that a stone outcrop overhanging the rock might have provided such a

Fig. 31. Deterioration of a marble rock art panel in Sagan-Zaba (Photo: Ekaterina Devlet) (left)

Fig. 32. Laminar and granular exfoliation and algae growth are the main causes of stone deterioration at Sagan-Zaba (Photo: Elena Miklashevich) (right) "roof" had it not come down centuries ago. Flaking of the surface caused by weathering and the threat of laminar exfoliation (to a depth of up to 10 cm) still exists.

From 1992 to 1996, in collaboration with E. Ageeva (Department of Restoration for Stone Monuments at the State Institute of Restoration) and N. Rebrikova (Dr. of Biology, Head of the Biological Laboratory in the same Institute), I examined six rock art sites in the Baikal area (Sagan-Zaba, Aia, Orso, Malaya Orso, Elgasur, Sachurte and Sarma) and assessed their state of preservation and the dynamics of the rock engravings' deterioration. We then drew up recommendations for their preservation and priorities for their protection and management.

We came to the conclusion that the deterioration was caused mainly by precipitation and direct water flow after the loss of natural protective "roofs" and ledges and by water percolation through cracks. To prevent further natural deterioration we made recommendations mainly involving indirect modifications (Ageeva et al., 1995; Ageeva, Devlet E. and Rebrikova 1996). Each summer we also visited two main sites (Sagan-Zaba, Orso) for monitoring purposes and found that in this area, as well as all over the world, rock art sites in remote places are in a better state of preservation than those with easy visitor access.

In 1994 at Orso we tried to reconstruct some of the previous conditions to be found at a rock art site (Figs. 33a, b). It consists of a small marble surface with engraved figures of deer dating from the medieval period. For its preservation we created water drip lines and changed the direction of water flow so as to prevent its direct contact with the decorated surface. Some areas outside the surface with engravings were treated with biocides to prevent further algae growth. We tried to confine this work to indirect modification of the site and reconstruction of the original conditions rather than repair, but some direct interference proved necessary.

The Orso rock art site in the vicinity of Lake Baikal was badly damaged by destructive natural processes. The loss of protective "roofs" projecting above the surface with petroglyphs resulted in laminar and mass-scale exfoliation and other biological deterioration. Rock art in the early-medieval style is pecked on a weathered marble surface measuring 1.5 x 1.7 meters and is covered with a

Figs. 33a, b. The Orso rock art site, Lake Baikal, Siberia. Conservation of the part above the surface with rock art to prevent water flow and protect the patinated layer on the surface (Photo: Elena Miklashevich)

yellow patina consisting of calcium carbonate and iron hydroxide. The surface with rock art is inclined at an angle of 12 degrees and this serves to prevent raindrops and moisture collecting and thus helps it to survive. In 1994 direct and indirect conservation work was carried out to protect the rock art. The main focus of the conservation project was to reconstruct conditions favorable to its preservation (Ageeva and Rebrikova 1999). A protective projection made from local stone fused with a silicone binder was introduced so as to prevent water flow. Above the decorated surface, the inclined part of the rock contained deep cracks and depressions led to the retention of snow, water, and various sediments, which all promoted further deterioration. In the area above the rock art cracks were filled with small fragments of stone and sealed with cement. At the edges of the decorated surface the surviving patinated crust was weathered. The exfoliated edges in the damaged areas were strengthened down to their sub-stratum, using slaked lime, grained marble and a silicone binder. The rock surface had been totally invaded by algae and lichen (Xanthera elegans, Phaeophyscia sciastra, Physcenia muscigena, Lecanora frustulosa, Caloplaca flavovirescens, Candelariella aurelia, Pseudophebe minuscula, Sarcogyne regularis, etc.) and it was treated with biocides in order to prevent further growth. The monitoring of the area undertaken since then revealed that the surface water-flow has been re-routed, the newly introduced materials are proving effective and the conservation materials are still water-resistant.

Rock Art Sites in the Middle Yenisei River Basin

Attracting the public to rock art sites is now an important activity in many regions of Russia. In Khakassia, a region in South Siberia, a good deal has been done to interest people in cultural heritage. Important museums were set up in Kasanovka and other regional centers. The main problem for these areas with highly attractive sites is too much emphasis on tourism and not enough on rock art protection. From 2009 to 2013 a special program was introduced aimed at developing tourism at Khakassia's rock art sites. Each local center was allocated funds for an open-air museum where rock art was the core attraction. The most important thing is a positive approach to the protection and control of rock art. Earlier attempts to conserve certain panels failed because of uncontrolled visitor access, which undermined all the efforts of archaeologists and conservation specialists (Figs. 34-37). The immense rock art sites at this location require professional help for their preservation. Regrettably, at many rock art sites the terrible consequences of uncontrolled visitor access to the rock outcrops, even to the ancient art, are clear for all to see (Esin 2011; Miklashevich 2003, 2008).

Fig. 34. Removal of grafitti, Middle Yenisei region, Siberia (Photo: Elena Miklashevich)

Fig. 35. Open-air rock art museum in Poltakovo displaying rock art and decorated stelae from local excavations in Khakassia, Siberia (Photo: Elena Miklashevich)

Fig. 36. Damage to rock art on steles in the open-air museum of an excavated burial-mound at Salbik in Khakassia, Siberia (Photo: Elena Miklashevich)

Fig. 37. A modern building in the form of a traditional dwelling at the Ulug Khurtuyakh-Tas Museum in Khakassia, Siberia (Photo: Yuri Esin from "Ulug Khurtuyakh Tas: The Cultic Stone of Khakassia with the History of Four Thousand Years." Abakan, 2011, p.15)

Re-location

Although it is highly controversial, another way to preserve images under threat of submersion is to re-locate boulders and panels with engravings to museums. For many sites threatened with flooding or physical destruction resulting from other factors, this is the only way to save the rock art for future generations. Opportunities for moving rock art safely are extremely limited but there are some examples of the successful re-location of rock art from Angara and the Middle and Upper Yenisei. In such cases, the rock art survived but lost its environmental and cultural context (Fig. 38). In some areas, re-located rock art and steles are displayed in open-air museums (Fig. 37). There are also cases of damage resulting from re-location: for example, when a huge engraved granite boulder at Onega was removed for re-location to Leningrad in 1935 using dynamite to separate the decorated surface from the surrounding rock, a unique panel depicting childbirth was destroyed.

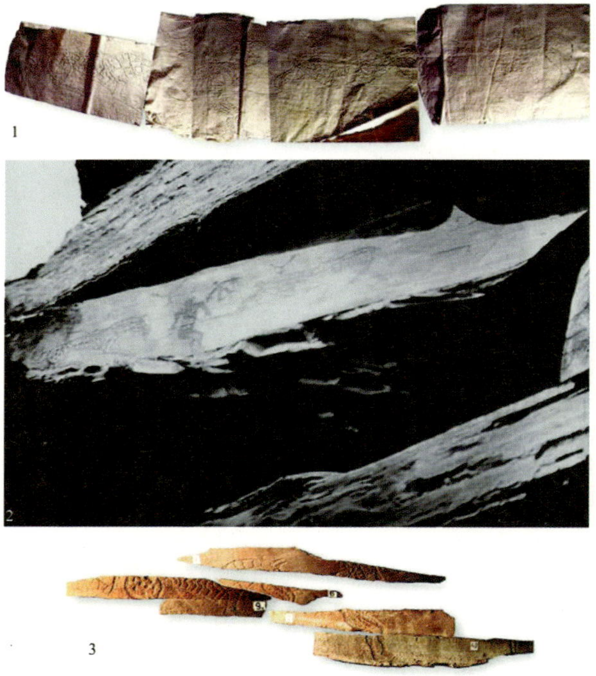

Fig. 38. A comparison of a lost rock art panel from Oglakhty in the Middle Yenisei region (Photo and rubbings of 1907: A. Adrianov); a fragment of the same panel from the Abakan Museum (Photo: Elena Miklashevich)

Conclusion

Several years ago rock art sites seemed to be an inexhaustible element of cultural heritage. Unfortunately, part of this heritage has been lost irrevocably. Numerous natural processes are involved in the destruction of rock art but the most disastrous damage has been caused by huge industrial projects. Rock art sites of great importance have been lost because of the construction of hydro-electric power-stations and reservoirs on the major rivers of Siberia. Outstanding sites have been damaged and many others have been partly or completely flooded. In such cases the recording of rock art images was the only way to preserve information for scientific research. Under such circumstances, before the introduction of 3D technologies, making casts, rubbings, and recordings were the only practical ways to preserve rock art (Figs. 39-41). Attempts were made to relocate some engraved rock boulders, which were under threat of submersion, to museums so as to preserve and display rock art in open-air collections.

Fig. 39. Cast of a rock art panel from Pegtymel, Chukotka (Photo: Evgenii Girya)

Fig. 40. Cast of a stone block with rock art from Pegtymel, Chukotka (Photo: Alexei Kochanovich)

Fig. 41. Exhibition of rock art rubbings from Besovye Sledki and Zalavruga in Kondopoga, Northwest Russia (Photo: Igor Georgievskii)

The main problems involved in rock art protection and management are similar throughout the vast territory of Russia. In recent years natural elements and direct anthropogenic impact have been responsible for changing environmental conditions, which have contributed significantly to the deterioration of rock art in Russia. Unfortunately, industrial activities and vandalism are also contributing greatly to rock art deterioration. I wish to avoid negative comments and accounts concerning rock art conservation projects of the last century. All these projects seem inadequate and irrelevant now, but they derived from the opportunities and understanding that was available then and were an attempt to do at least something, rather than merely observe unmanageable deterioration and destruction (Devlet 1999; Miklashevich 2003). A Siberian project first launched in the 1990s was successful because of the efforts of an inter-disciplinary team involving archaeologists, geologists, engineers, and conservators from chemical, physical, and biological laboratories of the State Institute of Restoration, who had wide experience in the conservation of stone in the open air. Different aspects of rock art studies and protection were investigated. The main factors contributing to deterioration were found to be degradation of rock, destabilization of slopes, degradation of patina and pigment loss, widespread biological decay, obscuring of art through deposits of salt and damage resulting from unregulated tourism and vandalism. The adoption of international approaches to rock art preservation slowly, step by step, has led to the recognition of it being a multi-disciplinary problem, involving questions of protection, legislation, public awareness, preservation and management strategies, direct and indirect conservation measures, pressure from development and tourism, air pollution, and environmental changes. Russian specialists in rock art protection and research also turned to experience gleaned during projects that had already achieved fruitful results and won international respect (Rosenfeld 1988; Pearson, C. and Swartz, B.K. 1991; Hygen 2011; Smith, Helskog and Morris. eds. 2012; Whitley. ed. 2001).

At some sites a great deal was done for the conservation of panels with rock art but the absence of regular controls resulted in a resumption of vandalism. The systematic control of visitors and the dissemination of knowledge regarding the meaning of rock art and its importance as a part of world cultural heritage are still essential for effective rock art conservation. I suggest that promotion of the public's knowledge and enjoyment of rock art has a crucial role to play in its protection. The existing contradictions between the demands of protection on the one hand and those of economic development and everyday use of territories containing rock art on the other are still acute. The development of tourism has opened up new possibilities for residents of areas with rock art but has also called into being new problems regarding the preservation of landscapes with rock art.

BIBLIOGRAPHY

Ageeva, E.N., E.I. Antonova, N.L. Rebrikova, and B.T. Sizov. 1993. "Shishkinskie Pisanitsy. Rezul'taty Obsledovaniya i Predlozheniya po ikh Sokhraneniyu," *Pamyatniki naskal'nogo iskusstva,* Moscow, pp. 25-29.

Ageeva, E.N., E.G. Devlet, and N.L. Rebrikova. 1996. "Resultati Obsledovania, Perspektivi Sohranenia i Ispolsovania Pamiatnilov Naskalnogo Iskusstva Ozera Baykal," *Arkheologisheskoye nasledie Baykalskoy Sibiri,* Vol. 1, Irkutsk, pp. 111-115 (In Russian).

Ageeva, E.N., E.G. Devlet, and N.L. Rebrikova and M.Ya. Sklyarevskiy. 1995. "Sostoyanie Pamyatnika Naskal'nogo Iskusstva Sagan-Zaba i Perspektivy ego Sokhraneniya i Ispol'zovaniya," *Naskal'noe iskusstvo Azii* 1, pp. 29-30.

Ageeva, E.N. and A.V. Kochanovich. 2011. "Preventivnaya Konservatsiya na Pamyatnike Naskal'nogo Iskusstva Tomskaya Pisanitsa," *Rock Art in Modern Society, On the 290th anniversary of the discovery of Tomskaya Pisanitsa, Book of papers of the International Conference,* Vol. 1. Kemerovo (Occasional SAPAR publication. Vol.VIII), pp. 172–175.

Ageeva, E.N. and N.L. Rebrikava. 1999. "Orso Rock Art Site in Siberia. Practice of Conservation," International Prehistoric Art Conference, August 3-8, 1998, *Proceedings,* Vol. 1. Kemerovo: NIKALS, pp. 254-256 (In Russian, with English summary).

Bahn, P.G., R.G. Bednarik and J. Steinbring. 1995. "The Peterborough Petroglyph Site: Reflections on Massive Intervention in Rock Art," *Rock Art Research* 12 (1), pp. 29-41.

_________. 1996. "A Canadian Tragedy," *International Newsletter on Rock Art* 13, pp. 14-17.

Bednarik, R.G. and K. Devlet. 1992. "Rock Art Conservation in Siberia," *Puracala* 3(1-2), pp. 3-11.

Devlet, E.G. 2008. "Rock Art Studies in Northern Russia and the Far East," *Rock art studies. News of the World* 3, Oxford: Oxbow, pp. 120-137.

_________. 2010. "New Developments in the Study of Rock Art of Northern Eurasia," *Northern Archaeological Congress III Papers.* November 8–13, 2010, Khanty-Mansiisk, Ekaterinburg-Khanty-Mansiisk, pp. 180-208.

_________. 2012. "Rock Art Studies in Northern Russia and the Far East," *Rock art studies. News of the World* 4, Oxford: Oxbow, pp. 124-148.

Devlet, E.G. and M.A. Devlet. 2002. "Heritage Protection and Rock Art Regions in Russia," *L'art Avant l'histoire. La Conservation de l'art Préhistorique. 10es Journées D'études de la Section Française de L'institut International de Conservation,* SFIIC, Paris, pp. 87-94.

_________. M.A. 2005. *Myths in Stone: World of Rock Art in Russia,* Moscow: Aletheia (In Russian and English).

Devlet, M.A. 1980. *Petroglifi Mugur-Sargola,* Moscow: Nauka (In Russian).

_________. 1998. *Petroglify na dne Sayanskogo Morya (gora Aldy-Mozaga),* Moscow: Pamyatniki istoricheskoi misli (In Russian).

Esin, Y., L. Gorbatov, 2011. Ulug Khuryuyakh Tas, Abakan.

Francfort, H.P. and Ja. A. Sher (sous la direction de). 1995. *Repertoire des Petroglyphes d'Asie Centrale, Fascicule N°2: Siberie du Sud 2: Tepsej I-III, Ust'-Tuba I-IV (Russie, Khakassie), (Memoires de la Mission Archeologique Francaise en Asie Centrale, vol. 2),* Paris: Diffusion de Boccard.

Helskog, K., N. Lobanova, and A.-S Hygen, 2008. "Protection and Presentation of Rock Art in the Republic of Karelia, The Russian Federation," *The Norwegian-Russian Co-operation 1995-2008,* Oslo, Riksantikvaren, pp. 94-108.

Hygen, A.-S. 2011. "Management of Large Rock Art Landscapes: The Cases of Tamgaly (Kazakhstan), Sarmishsay (Uzbekistan), and Gobustan (Azerbaijan)–Experiences from a Norwegian Point of View," *Rock Art in Modern Society, On the 290th anniversary of the discovery of Tomskaya Pisanitsa, Book of papers of the International Conference,* Vol. 1, Kemerovo (Occasional SAPAR publication. Vol.VIII), pp. 39-47.

Kochanovich, A.V. 2005. "Issledovanie i Monitoring Sokhrannosti Pamyatnika Naskalnogo Iskusstva «Malaya Boyarskaya Pisanitsa»," *Problemi nauchnoi restavratsii pamyatnikov istorii i kulturi,* Moscow (In Russian).

Laskin, A.R. 2007. "Perspektivy Dal'neyshego Izucheniya i Sokhraneniya Petroglifov Sikachi-Alyana," *Arkheologiya, etnografiya i antropologiya Evrazii,* 2, pp. 136-142.

_________. 2009. "The Rock Art of Sikachi-Alyan: Future Study and Preservation," *Archaeology, Ethnology and Anthropology of Eurasia, Issues in the Study of Prehistoric Art, Discussion,* Novosibirsk, pp. 262-268.

Laskin, A.R., E.G. Devlet, A.L. Babaev, and A.I. Sudakov. 2005. "Petroglify Sikachi-Alyana-unikal'nyy Pamyatnik Drevnego Naskal'nogo Iskusstva na Nizhnem Amure Problemy Sokhraneniya i Ispol'zovaniya," *Mir naskal'nogo iskusstva,* Moscow, pp. 154-162 (In Russian).

Lobanova, 2009. "Patroglyphs at Staraya Zalavruga: New Evidence - New Outlook," *Archaeology, Ethnology and Anthropology of Eurasia, Issues in the Study of Prehistoric Art, Discussion,* Novosibirsk, pp. 224-232.

Martynov, A.I. 1988. *Pisanitzi na Tomy,* Kemerovo (In Russian).

Martynov, A.I. and G.S. Martynova. 1993. "The Tom Drawings: An Ancient Sanctuary in Nothern Asia," *International Newsletter on Rock Art 7,* pp. 18-21.

Medvedev, V.E. 2011. "On the History of Organizational-protective Measures for the Petroglyphs," *Rock Art in Modern Society, On the 290th anniversary of the discovery of Tomskaya Pisanitsa, Book of papers of the International Conference,* Vol. 1, Kemerovo (Occasional SAPAR publication. Vol.VIII), pp. 179-183.

Melnikova, L.V., V.S. Nikolaev, and N.I. Demyanovich. 2011. *Shishkinskaya Pisanitsa,* Vol. 1. *Istoria i metodika izuchenia, problemi muzeefikatsii, opisanie petroglifov,* Irkutsk: IZK (In Russian).

Miklashevich, E. A. 2003. "Rock Art Research in North and Central Asia 1995–1999," *Rock Art Studies News of the World,* II, Oxford: Oxbow Books, pp. 88-118.

_________. 2008. "Rock Art Research in Siberia and Central Asia, 2000-2004," *Rock Art Studies News of the World,* III, Oxford: Oxbow Books, 2008, pp. 138-178.

_________. 2011. "Documentation of Damage to the Petroglyphs of Tomskaya Pisanitsa," *Rock Art in Modern Society, On the 290th anniversary of the discovery of Tomskaya Pisanitsa, Book of papers of the International Conference,* Vol. 1, Kemerovo (Occasional SAPAR publication. Vol.VIII), pp. 128-137.

_________. 2011. "Dokumentirovanie Povrezhdenia na Pamiatnike Naskal'nogo Iskusstva v Bukhte Sagan-Zaba," *Drevnie Kul'tury Mongolii i Baikal'skoi Sibiri,* Irkutsk: Irkutsk State Technical University, pp. 532-540 (In Russian).

Okladnikov, A.P. 1959. *Shishkinskie Pisanitzi,* Irkutsk: Irkutskoe knizhnoe izdatelstvo (In Russian).

_________. 1974. *Petroglifi Baykala-Pamiatniky Drevney Kulturi Narodov*

Sibiri, Novosibirsk: Nauka (In Russian).

________. 1977. *Petroglifi Verkhnei Leni,* Leningrad: Nauka (In Russian).

________. 1981. *Ancient Art of the Amur Region,* Leningrad: Aurora Art Publishers.

Okladnikov, A.P. and Martynov, A.I. 1972. *Sokrovisha Tomskih Pisanits,* Moscow: Iskusstvo (In Russian).

Pearson, C. and B.K. Swartz. 1991. *Rock Art and Posterity: Conserving, Managing and Recording Rock Art, Occasional AURA publication* 4, Australian Rock Art Research Association, Melbourne.

Perjakova, T. and M. Sklyarevski. 1997. "From the Experience of the Center of Preservation of Irkutsk Historical and Cultural Heritage," *International Newsletter on Rock Art* 17, pp. 16-18.

Podovkina, E.A. and O.I. Goriunova,1995. "Pisanitzi Sarminskogo Ushelia (oz. Baykal)," *Kulturi i Pamiatniki Bronzovogo i Zheleznogo Vekov Zabaikalia i Mongolii,* Ulan-Ude: Izdatelstvo Buryatskogo nauchnogo tsentra, pp. 119-122 (In Russian).

Rosenfeld, A. 1988. "Rock Art Conservation in Australia," *Special Australian Heritage Publication.* Series 2. Canberra: AGPS.

Savvateyev, Yu. A. 1990. *Kamennaia Letopis Karelii. Petroglifi Onezhskogo Ozera i Belogo Moria,* Petrozavodsk: Karelia (In Russian).

Savvateyev Y. 1994. "The White Sea Rock Engravings," *International newsletters on rock art 7,* pp. 22-27.

Sher, J. 1999. *Siberie du Sud* 4: Cheremushny Log, Ust'-Kulog, Repertoire des Petroglyphes d'Asie Centrale, Fascicule 4, Paris: De Boccard, pp. 5-46.

Sher, J., N. Blednova, N. Legchilo, and D. Smirnov. 1994. *Siberie du Sud* 1: Oglakhty I-III (Russie, Khakassie). Repertoire des petroglyphes D'Asie Centrale, Fascicule N 1, Paris.

Smith, B.W., K. Helskog, and D. Morris. 2012. *Working with Rock Art: Recording, presenting and understanding rock art using indigenous knowledge,* Johannesburg.

Shumkin, V.Ya. and E.M. Kolpakov. 2012. *Rock Carvings of Kanozero.* St.Petersburg: Art of Russia, p. 424. ill.

Soleilhavoup, F. 1994. "The Conservation of Rock Art and Land Management: Contradictions and Paradoxes," *International Newsletter on Rock Art* 7, pp. 11-15.

Wainwright Ian, N.M. 1985. "Rock Art Conservation in Canada," *Bolletino del Centro Camuno di Studi Preistorici* 22, Capo di Ponte.

Whitley, D.S. 2001. *Handbook of Rock Art Research, Walnut Creek,* Lanham, New York, Oxford.

07

The State of Rock Art Management and Conservation in North America

Johannes H. N. Loubser

The vast majority of rock engravings (petroglyphs) and rock paintings (pictographs) in North America are the work of Native American Indians (referred to as Indians in most of the United States and as First Nations in Canada), but a substantial number of historic period Euro-American engravings and paintings are also to be seen. Whereas rock art sites are particularly prevalent in the arid regions west of the Mississippi River (Wellmann 1979), a surprising number of new sites have been discovered recently in the comparatively wet eastern woodlands (Diaz-Granados and Duncan 2004). With an increasing awareness of rock art among culture heritage tourists (National Trust for Historic Preservation 2011), archaeologists (Sánchez et al. 2008), bureaucrats (Keyser 2005), and the general public (Whitley 2004), the need to protect rock art sites in North America has received increasing attention since at least the late 1980s. Along with an increasing interest in rock art sites and their appropriate management has been an increase in fieldwork to locate, record, research, and manage the sites. The passing of the United States Native American Graves and Repatriation Act (NAGPRA) in 1990 further bolstered concern for the proper treatment of Indian cultural items, including rock art sites, which, among other pro-active measures, necessitated consultation with relevant indigenous groups.

This chapter reviews the management and conservation of three different rock art sites in the United States of America and Canada. Whereas the management and conservation in these countries have been influenced by developments in France and Australia (e.g., Achiary et al. 1995), substantial knowledge was shared with archaeologists and rock art researchers during a series of workshops presented by the Getty Conservation Institute in the late 1980s and early 1990s (e.g., Stanley-Price 1994). Compared to elaborate infrastructural changes made at a number of French and Australian rock art sites, changes to accommodate visitors at rock sites in Canada and particularly the United States have been comparatively minimal. As will become apparent in this chapter, evidence suggests that over the long term a minimal infrastructure at rock art sites might yet prove to be the most sustainable option.

The main aim of this chapter is to show that no one single management and conservation strategy fits all sites, primarily due to the unique sets of natural, cultural, historical, and managerial circumstances associated with each site. A well thought-out and systematically set-out management plan, based on thorough background research, meticulous field observations, and consultation with concerned and knowledgeable stakeholders, is a prerequisite for a properly managed and conserved rock art site.

Rock Art Site Management as an Inevitable and Ongoing Process

Being guided by the wishes of concerned and informed stakeholders, caretakers have to pro-actively manage rock art sites in an ordered fashion or otherwise uninformed visitors will set the management standards, often in a chaotic fashion and to the detriment of the site and its art. Given that a site is in a good natural condition, good site preservation is really all about good presentation to the visiting public. For example, if a site has graffiti and appears untidy and uncared for, then visitors cannot take all the blame for not taking care of or for behaving inappropriately at a rock art site. Clean rock surfaces devoid of dust and clean deposits without litter or detritus create a good impression. The presence of minimal signage and perhaps a walking surface create a sense of official presence and custodial care.

To determine which management and conservation strategy best fits a particular site, it is necessary to conduct the following step-like investigation:

> 1) *background* literature synthesis and interviews with knowledgeable people concerning environmental and historical settings; 2) mapping, photography, tracing, and digital enhancements of the surrounding terrain, the site, and its rock art in the field and office to obtain *a baseline graphic record;* 3) a thorough assessment of a *site's significance* by investigating its spiritual, historical, tourist, aesthetic, and research values; 4) a systematic assessment of a site's condition by closely examining its rock support, motifs, surface accretions, associated ground deposits, and terrain surrounding the site; 5) an assessment of a *site's management* by investigating the ownership history and current opportunities and constraints; 6) based on a site's significance, condition, and management, it becomes possible to recommend conservation, visitor and marketing strategies that are minimal, repeatable, compatible, and distinguishable from the current one; and 7) always keep in mind the necessity of monitoring, re-assessment, and possible modification of management and conservation recommendations.

Depending on which kinds of significance value predominate and how they combine with specific site condition and management contexts, rock art sites in North America are normally managed in one of the following three basic fashions:

> 1) as *focal points* for visitation (normally where tourist significance predominates, robust site conditions prevail, traditional stake-holders agree, and there is a well-established management infrastructure); 2) as *low-level heritage sites* (normally where spiritual significance predominates, robust site conditions prevail, and the management infrastructure is part-time) or by 3) *closing sites* from public visitation (normally where research or spiritual significances predominate, site conditions are friable or constricted, the site has a long history of vandalism, and management infrastructure is part-time). Of these three options, closing sites from public visitation is the most difficult

> to enforce, especially when people already know where the sites are located. Opening sites to the public in a well-managed fashion has many advantages, the most important being that visitors are educated and instilled with a renewed sense of appreciation of rock art and the need for its ongoing custodial care. Well-interpreted rock art sites teach visitors their importance, and by extension, the benefits of preserving similar places elsewhere.

It is not the presence or quantity of visitors at sites that are necessarily deleterious but rather their uninformed notions and resulting inappropriate behavior, such as touching or scratching the rock art. Moreover, educated people realizing the significance of rock art go a long way to police sites against vandals in the absence of official agency personnel. It is important to realize that sites opened, managed, and interpreted specifically with public visitation in mind often survive longer than those deliberately closed to visitors or those left without any management plan.

Appropriate management practice, as understood by current conservators, is to slow the clock of natural decay and human damage in order to preserve the integrity of a rock art site for future generations to enjoy, study, and conserve. To this end, any management decision should proceed in such a fashion as to have minimum impact on places with rock art, most notably by keeping infrastructural developments to a minimum. Managers, moreover, realistically aim at repeatable actions where, for example, future attempts at installing infrastructure should not be jeopardized by the choice of current materials and techniques, such as pouring hard-to-remove concrete pathways. Another concept to bear in mind regarding management infrastructure is compatibility. The use of compatible materials at places with rock art, for example, implies that relatively inert materials, such as wood or stone, should be preferred over corrosive materials, such as metal or cement. A fourth concept that applies to acceptable management practice is distinguishability. At any rock art site being managed, it is important to distinguish between what has been added from that which existed prior to intervention. For instance, any additional features that were not in the rock art site before, such as a newly constructed pathway, should contain modern chemical or physical markings that are not only compatible with the rock, but also distinguishable from the original rock.

To illustrate the points raised above, the history and current status of management and conservation are summarized for the following three sites: Writing-on-Stone (a Focal Point Site); Judaculla Rock (a Low-Level Heritage Tourist Site); and Paint Rock (a Closed Site).

Focal Point Site

Writing-on-Stone is a discontinuous line of sandstone cliffs, on average 30 feet high, overlooking the northern banks of the Milk River in southern Alberta, Canada. Located within the Writing-on-Stone Provincial Park, the cliffs contain over 50 petroglyph and pictograph site complexes and thousands of panels. The extensive Writing-on-Stone rock art complex is significant within Alberta and beyond as it contains an unusually varied collection of Plains Indian tradition petroglyph and pictograph panels, often depicting people and animals in minute detail. Petroglyphs include parallel vertical striated grooves, fully pecked figures, Ceremonial Tradition thin-line incised figures with shields and bison, and Biographic Tradition thin-line incised figures with muskets and horses. Pictographs include red ochre smears, painted and finger-applied red ochre shield-carrying figures, and painted animal motifs. In a few instances carefully excised petroglyph surfaces are in-filled with red ochre pigment.

The poor bonding-properties of the clay matrix within the sandstone are responsible for the friable nature of the rock surfaces in most locations, making the outer layer of Writing-on-Stone particularly susceptible to wind erosion. In situ weathering, such as caused by salts, appears to be less of a problem. What adds to the structural instability of the rock along bedding planes and joints are fluctuating seasonal and daily temperatures on the expansive and exposed grassy plains surrounding the rock art complex.

Known as Áísínai'pi (literally "it is pictured/written") by the Blackfoot Nation, the rock art and site complex is historically associated with these Plains group of gatherers and hunters. Blackfoot who made petroglyphs and pictographs at the site complex did so to depict a vision received in a dream from a highly individualistic and secretive dream helper (Dempsey 1973:23). The Blackfoot people believed that powerful seers among them could consult the images on the rock surfaces not only to find bison, enemies, or lost property, but also to foretell future events. Blackfoot informants have referred to the rock images as mirrors, where people can see their fate; altered or new pictures that appear overnight are believed to change events (Klassen 2007:17). The last documented Blackfoot raiding party to consult the rock art at Áísínai'pi was in 1889. Nonetheless, traditional Blackfoot beliefs revolving around the place persist to this day. For instance, the Blackfoot Indians still consider the area within and around the petroglyph complex as imbued with overwhelming spiritual potency (e.g., Brink 2007:63). The sacred significance of the rock art complex cannot exist without the Sweetgrass Hills that occur on the horizon to the south (Klassen 2007:20); alteration of the view shed from the complex will negatively affect the overall spiritual integrity of the entire landscape.

The Writing-on-Stone rock art complex is also significant from an archaeological research point of view, which explains for instance why archaeologists from the Alberta Provincial Parks Department surveyed and catalogued numerous sites in the 1970s. In preparation for a proposed fence installation, Brink (1979) excavated deposits beneath selected rock art panels in 1977. Although no big prehistoric camps were found, bone tools recovered during excavations could have been used to incise at least some of the petroglyphs.

The historic period in the park is also of archaeological significance, including the North-West Mounted Police tent camp dating back to 1887 and a permanent outpost that existed between 1889 and 1918. Considering that these sites are south of the Milk River, they are not directly affected by what is done at the rock art sites open to public visitation north of the river. Care should be taken nonetheless when removing historic period graffiti north of the river in case they may be associated with the North-West Mounted Police presence in the area. Care should also be taken not to remove names and dates left by early Euro-Canadian homesteaders who farmed the area, mostly between 1905 and the 1930s. For sentimental reasons, at least some descendants of these pioneer farmers and ranchers still visit the cliffs to view the names of their ancestors.

As early as the 1920s increasing numbers of people from farther afield started visiting the area for its rock art and natural scenery. Provincial authorities and citizens in Alberta have for some time been concerned about graffiti at Áísínai'pi (Klassen 2007), particularly after the Writing-on-Stone Provincial Park was created in 1957 without the necessary infrastructure. With the drastic increase in uncontrolled outside visitation in 1957, graffiti skyrocketed (Fig. 1). To better control and channel the movements of park visitors, the Archaeological Preserve was established in the western section of the park in 1977 and was set aside for guided tours. From then on entry into the gated Preserve was limited to those visitors who paid to join guided tours. Since designation of the Archaeological Preserve in 1977, graffiti declined drastically. In 1981, a portion of the park was named a Provincial Historic Resource to protect the rest of the rock art from increasing impact from vandalism and graffiti. In March 2005, the park was designated a National Historic Site. On June 20, 2007 the park's new visitor center, with views of the valley from the north rim, was officially opened. During this period of infrastructural development, most of the rock art panels were photographed and traced, creating a useful, interpretive and comparative base-line record for rock art scholars and conservators.

Within the Archaeological Preserve, a trail follows almost the entire line of cliffs. At least eight level areas have been constructed along the trail, each with benches to enable guided tour groups to congregate and properly view

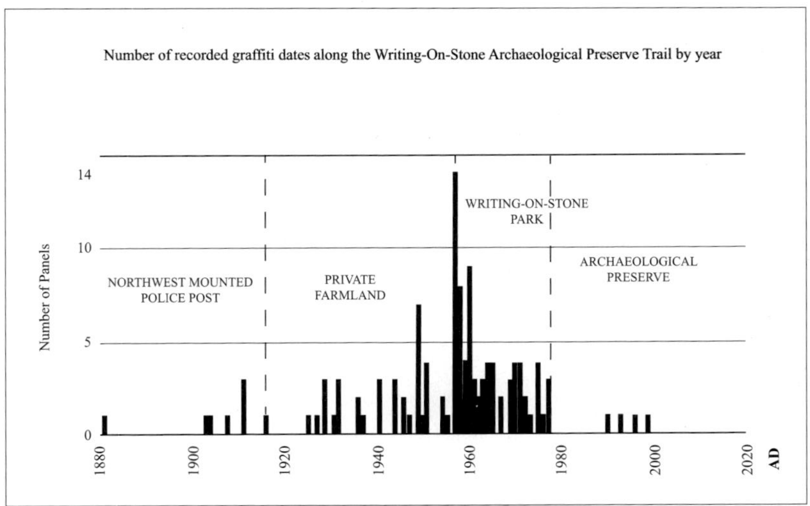

Fig. 1. Bar graph showing an explosion of graffiti at the establishing of the Writing-On-Stone Park and a marked decrease after the establishment of the Archaeological Preserve.

the petroglyphs and pictographs without causing erosion to the soil deposits along the base of the cliff line. Having guided interpretive tours to a select number of high quality rock art panels, is among the most effective ways of managing and protecting rock art sites. Bus-loads of visitors from the visitor center are shown the petroglyphs by trained and experienced guides, most of them Blackfoot people, who have an in-depth knowledge and a culturally-transmitted understanding of the petroglyphs and pictographs. No known rock art site management strategy can surpass this in terms of inculcating understanding and respect of the rock art and so help reduce the incidences of vandalism. The use of indigenous guides sharing an emic perspective with visitors is a unique example of excellent site presentation for other public focal point sites to follow.

The rock art complex is a National Historic Site that has also been placed on Canada's tentative list for World Heritage inscription, but such a nomination has not yet been submitted to UNESCO, mainly due to the extensive historic period graffiti that negatively impacts the integrity of the site. The Alberta Tourism Parks and Recreation, with the assistance of the Royal Alberta Museum, accordingly decided that graffiti be removed from select rock surfaces within Writing-On-Stone-Provincial Park. Graffiti was removed and camouflaged for the following reasons (Loubser 2013): first, graffiti distracts from the spiritual significance that the petroglyphs and pictographs have for the Blackfoot First Nation; second, graffiti begets more graffiti; third, dense concentrations of graffiti make it difficult to locate examples of the original rock art; fourth, deeply incised graffiti can facilitate preferential weathering; and finally, graffiti distracts from nearby rock art motifs and the general integrity of the rock art site complex.

In the summer of 2012, Loubser (2013) and a small crew of volunteers removed graffiti from at least 100 separate panels. To prevent graffiti from recurring in treated areas, Loubser recommended that a physical distance or artificial physical barriers be created between pedestrian and rock art panels. Physical distance can be created by re-vegetating paths close to faint petroglyphs and pictographs, while psychological barriers close to highly visible petroglyphs and pictographs can include interpretive pedestals and low fencing.

Ongoing efforts by park staff to involve all stakeholders, including First

Nation people and local farmers, in decision-making regarding the rock art at this focal point site is a textbook example of consultation and empowerment. The continued employment of First Nation people as guides to the rock art within the Archaeological Preserve not only gives voice to descendants of the original artists but also authenticity to interpretations. Also, by focusing visitor attention on the rock art within the Archaeological Preserve, numerous rock art sites scattered elsewhere in the area are rendered ostensibly invisible.

Low-Level Heritage Tourist Site

Unlike Writing-on-Stone where rock art is the focal point for visitors, people who visit Judaculla Rock do so mainly during their travels to view the scenery and bird life that the surrounding area has to offer. Visitors also like to hike, bike, fish, and camp in the mountainous area of western North Carolina. Located within a scenic valley, Judaculla Rock is a big curvilinear-shaped outcrop of soapstone with quarry scars and petroglyphs. It occurs on land owned by Jackson County, east of Caney Fork Creek, a major branch of the northwestward-trending Tuckasegee River. The petroglyph boulder occurs within an artificially created bowl-shaped depression, which is currently covered with mowed grass (previously a corn field) and bordered on the west by a thicket of river cane. The surface of the westward-slanting petroglyph boulder, which measures roughly 22 square meters, includes scars left by soapstone bowl extraction and numerous pecked petroglyph designs. The densely packed nature of the motifs, especially along the upper two-thirds of the rock, in many instances makes it difficult to distinguish between motifs. Nonetheless, the following motifs have been identified: cupules; curvilinear units; bowl-shaped depressions; stick figures; rills; concentric rings; curvilinear motifs; deer tracks; claw-like imprints; an arc; a cross-in-circle motif; and a winged shape. Petroglyphs that occur within three of the hollow scallops suggest that the petroglyph production post-dates soapstone bowl quarrying at the site, a finding that is supported by similar overlaps at smaller soapstone boulders in western North Carolina and northern Georgia.

Albeit relatively soft rock, the soapstone and pecked surfaces of Judaculla Rock are in remarkably good condition (Loubser and Frink 2008). The climate of the southeastern woodlands is generally humid, with fairly high summer day-time temperatures giving way to freezing conditions in the winter months. The only real natural threat to the rock is the growth and expansion of lichen, facilitated by shade below a dense canopy of tree branches.

Judaculla Rock gets its name from Judaculla, the Anglicized pronunciation of the Cherokee name Tsul'kălū' (Mooney 1900:477), or Tsu-tla-ka-la and even Jooth-cullah in the phonetic form of the Cherokee syllabary (pamphlet produced by Cherokee Studies at Western Carolina University 2007). According to Mooney (1900:477) the name Tsul'kălū' means "he has them slanting," being understood referring to the pupils of the giant's eyes. In its plural form, Judaculla is the name given to giant spirit-beings who live in the west, also known to be the land of the dead. This use of the name in a slightly different context and the fact that Judaculla is mentioned as early as 1823 by Haywood, who spells the name as Tuli-cula, show that belief in this deity has antiquity.

Judaculla had dominion over game animals, many of which he hid in his extensive underworld abode, which included places such as Tannasee Bald, some 10 miles to the east of Judaculla Rock (e.g., Mooney 1900:262). To appease Judaculla as the Master-of-Game and so ensure success in hunting, some Cherokees tried to visit him at remote places such as Tannasee Bald, while others included his name in their formulas. Among other things, the boulder and its petroglyphs clearly served as a stopping point, or prayer location, marking the entrance into Judaculla's upland domain from the valley below. Even after their mass removal from the area in 1838, many Cherokees as far afield as Oklahoma continued to visit the rock on an annual basis for ceremonial purposes (Wilburn 1952:21).

The area near the petroglyph boulder has been used for a variety of purposes by Euro-Americans since the 19th century, including lumbering, mining, pasture for cattle, corn fields, and chicken houses. Milas Parker and his son, J. B. Parker, farmed the area and looked after Judaculla Rock until J. B. Parker donated the petroglyph boulder and surrounding land to Jackson County in 1959 (Parker 2006:12).

Although the Jackson County Board of Commissioners had good intentions to protect Judaculla Rock, their infrastructural changes to the landscape around the rock had some negative results. Backfill dirt from the upgraded and leveled road above the site eroded downslope and was redeposited around the petroglyph boulder. A cinder block building that was built around the rock in 1962 resulted in overcrowding and graffiti within the dark structure, accompanied by trampling of sandy mud onto the rock surface during rainy weather. Rainwater dripping from the roof also created a furrow around the structure. In order to rectify these problems the county dismantled the structure in 1966. The building was replaced with an open-plan roofed gazebo structure. Due to the continuation of silting and people picnicking on the rock, the county removed the gazebo in 1998 and replaced it with a wooden walkway and interpretive signs.

In 2006, at the instigation of the North Carolina Rock Art Project, members of the Caney Fork community formed a joint committee with Jackson County officials and representatives from the Eastern Band of Cherokees to solve the silting up of Judaculla Rock (Goble 2007). The committee decided to fund the thorough recording, condition assessment, and controlled archaeological excavations of deposit around the petroglyph boulder (Loubser and Frink 2008). Assessment of the excavated soils and comparison with old photographs showed that the average rate of annual soil accumulation was 2.7 centimeters. At this rate, the boulder with its petroglyphs would eventually have been completely covered by soil. As a result of intensive recording and condition assessment the following conservation and management actions were recommended for Judaculla Rock: re-direction of surface water from the road; removal of soil around the rock to the 1920s level; installation of an elevated walking surface down slope from the rock; installation of a wheel-chair accessible trail; removal and trimming of trees that block sunlight from eliminating lichen; installation of graphic interpretive panels; use of Cherokee perspective of the rock; installation of signs specifying visitation hours, parking, and boundaries; and installation of a visitor's book in a pedestal box.

Given the significant depth of the historic period soil disturbances, it was recommended that proposed infrastructural improvements at the Judaculla Rock site will not adversely affect any prehistoric archaeological remains (Shumate and Loubser 2011). Jackson County contracted a professional landscaping and design firm to install a semi-circular elevated viewing platform, assembled from sections of very durable precast slabs (Fig. 2). Graphic interpretive signs were installed along the handrails by professional

Fig. 2. The elevated walkway at Judaculla Rock soon after its installation

contractors, identifying and interpreting pecked features on the rock to visitors in terms of Cherokee perspectives.

Considering the managerial mistakes made at Judaculla Rock in the past, the fairly good condition of the petroglyph boulder is remarkable, which is testimony of the soapstone's resilience. Fortunately, the more recent collaborative pro-active and hands-on conservation and management of Judaculla Rock by various stakeholders have turned the site into a textbook example of how rock art sites in other parts of the country can be preserved, interpreted, and gainfully presented on a low-level but sustainable basis to the visiting public.

Closed Site

Due to the risk of damage and injury to rock art and visitors, sites with limited room and light within are normally closed to public visitation (e.g., Loubser and Boszhardt 2004; Raymond 2008). Rock art sites that are friable, pose a physical danger to visitors, or have a history of extreme and repeated abuse also warrant to be made invisible to the public eye. A site that has been subjected to repeated vandalism is Paint Rock, an open air-cliff face immediately east of the

French Broad River near the North Carolina-Tennessee state line.

The metamorphic sandy meta-limestone cliffs onto which prehistoric red and yellow pigments have been applied, some eight meters above ground level, appear to be in a reasonably good condition. Considering the wet and exposed conditions in the cliff, together with marked seasonal shifts in temperature, it is remarkable that the pigment has survived intact on the rock. The south-facing aspect of the panel receives sun most of the day, a factor that explains the well-developed case-hardened surface crust and silica skins. Consultation of historic sources shows that numerous painted panels once occurred against the cliffs, of which only one survived undamaged.

As early as 1790 there was a report of red paintings at Paint Rock, which included iconographic motifs such as humans, animals, fish, and birds (referred to by Cambron and Waters 1995:168). Unfortunately, by 1799 these were covered with black stains, reputedly resulting from the smoke from fires at the base of the cliff. The immediate area surrounding Paint Rock has been heavily unitized and traveled in historic times. The prominent location of the painted panel, above an old trail, meant that the locale was used for a variety of purposes, including a defensive post during skirmishes with the Cherokees (Ramsey 1859:569-570), a stage coach line (Colton 1859:135), a toll road and possible toll booth, a resting spot where people like to write on the rock (Zeigler and Grosscup 1883:370), and a place where drovers herded animals to eastern markets (Jefferson 1998). More recently picnickers and revelers have used and abused the main overhang portion of the site, such as leaving trash, making fires, and adding graffiti to the rock face.

In spite of all these deleterious activities, a bichrome red and yellow panel on the North Carolina side of the state line has survived remarkably intact. It is this panel that was carefully recorded in 2006 (Loubser 2007) (Fig. 3). Small painted flakes were collected for Energy Dispersive Spectrometry and Accelerator Mass Spectrometry. The dating and physical analysis results were unexpected for a number of reasons. First, the likely 5,000-year-old date for the rock art goes against prevailing beliefs that rock paintings in the Southeastern United States must be young due to adverse climatic conditions. Secondly, the amount of carbon within the rock was far higher than is typically the case. Thirdly, comparatively high levels of sulfur in the yellow pigment suggest that this pigment came from a different source than the red. Not only does the rock

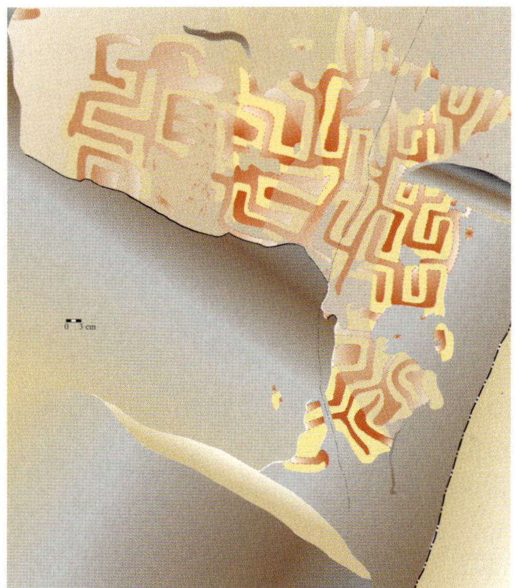

Fig. 3. Re-drawn tracing of the red and yellow panel at Paint Rock

art appear to be older than expected, but the efforts that went into obtaining and applying the pigment were remarkably involved too.

Also remarkable is that only a handful of people know about the prehistoric paintings high up the cliff face. The Cherokees are not aware of the Paint Rock panel either, possibly due to its great antiquity. It is likely its invisibility that has saved the painted panel from smoke stains and graffiti that are covering most of the other rock surfaces. The considerable research potential of the only intact painting at Paint Rock highlights the need for its conservation and favors the decision that its location is kept a secret. Visitors cannot be stopped from going to the well-known spot, but management measures can be implemented to minimize the chances of damage being done to the paintings. One measure to discourage visitors from approaching the cliff below the painted panel is to revegetate it with thorny shrubs. Careful removal of soot and graffiti from the northern portion of the site very likely will reveal the other paintings that have been covered since many years ago. Erecting visible fences or installing a gate will only aggravate matters, drawing unnecessary attention and challenging people to break in. "Closing" the site in this instance is simply maintaining the ostensible invisibility of the only intact panel and making it difficult for visitors to reach the cliff directly below.

Concluding Remarks

A common trend at all three sites discussed is that graffiti does not necessarily increase with increasing population or increasing visitation numbers. A shared reason for graffiti rather appears to be unmanaged sites, be it through opening them to the public without any prior planning or constructing roads to within easy walking distance of the sites. Re-routing of roads, limiting entrance to guided tours only, and increasing monitoring by park staff or trained site stewards have been accompanied by a rapid drop-off in graffiti incidences. But above all, effective presentation of appropriate interpretive information through trained guides and visual media is perhaps the most successful deterrent. Positive signage, such as "please help us preserve these rare rock markings by staying on the pathway," also goes a long way to help foster a feeling of custodial care among visitors.

Surviving portions of the many rock art sites that have been damaged in North America due to actions by people who were unaware of their significance are increasingly being managed and conserved through well-recorded, researched, and planned actions. Agencies and land-owners are becoming increasingly aware of the benefits involving trained and experienced archaeologists specializing in the identification, recording, assessing, interpretation, and conservation of rock art sites (Copeland 2008). For sites to appear meaningful to a visiting public, it is necessary to create an awareness of the range of their significance values and incredible unrealized potential. An increased awareness and potential can only come to fruition when specialists are involved in the ongoing process of management and conservation. Also, it is imperative that descendants of the artists and stakeholders from the surrounding communities be involved in consultation and decision-making. Often it is not only the rock art site and its immediate surroundings that need to be preserved, but also the surrounding landscape and view which once were tied in with the rock art.

BIBLIOGRAPHY

Achiary, P., J. Brunet, B. Laurent, and P. Vidal. 1995. "A New Management Strategy for Cave Sites: The Sare Grotto," *Management of Rock Imagery,* G. K. Ward and L. A. Ward, eds, Melbourne: Occasional AURA Publication 9, pp. 13-19.

Brink, J. W. 1979. "Archaeological Investigations at Writing-on-Stone," *Archaeology in Southern Alberta,* Edmonton: Archaeological Survey of Alberta, Occasional Paper 12(13), pp. 1-74.

_________. 2007. "Rock Art Conservation Research at Writing-on-Stone Provincial Park, Alberta," *Journal of American Archaeology* 25, pp. 55-99.

Cambron, J. W. and S. A. Waters. 1995. "Petroglyphs and Pictographs in the Tennessee Valley and Surrounding Area," *Journal of Alabama Archaeology* 5(2), pp. 52-53.

Cherokee Studies at Western Carolina University. 2007. *People of the Land: An Introduction to the Cherokee Heritage Walking and Fitness Trail,* Cullowhee: Interpretive Pamphlet Published by the Western Carolina University.

Colton, H. E. 1859. *Mountain Scenery.* Raleigh: W. T. Tomeroy.

Copeland, J. M. 2008. "From Curiosity to Cultural Treasure: The Rock Art of Dinétah," *Set in Stone: A Binational Workshop on Petroglyph Management in the United States and Mexico.* J. P. Sánchez, A. Sánchez-Clark, and E. L. Abreu, eds, Albuquerque: Petroglyph National Monument, National Park Service, U.S. Department of Interior, pp. 67-81.

Dempsey, H. A. 1973. *A History of Writing-on-Stone,* Edmonton: Unpublished Manuscript on File at Alberta Recreation and Parks and the Provincial Museum of Alberta.

Diaz-Granados, C. and J. R. Duncan. 2004. eds, *The Rock-Art of Eastern North America: Capturing Images and Insights,* Tuscaloosa: The University of Alabama Press.

Goble, J. 2007. Community Members Share Ideas to Preserve Judaculla Rock, *The Sylva Herald and Ruralite,* March 29.

Haywood, J. 1823. The Natural and Aboriginal History of Tennessee, *Up to the First Settlements therein by the White People, in the Year 1768,* Nashville: George Wilson.

Jefferson, N. D. 1989. The Cultural Resource Survey for: Paint Creek Road Paving Nolichucky/Unaka Ranger District, Compartments 214, 217-219, Cherokee National Forest, Greene County, Tennessee.

Keyser, J. D. 2005. "Leader of the Pack: Government Sponsorship of Rock Art Research," *Discovering North American Rock Art.* L. L. Loendorf, C. Chippindale, and D. S. Whitley, eds, Tucson: The University of Arizona Press, pp. 217-237.

Klassen, M. A. 2005. "Aisinai'pi (Writing-on-Stone) in Traditional, Anthropological, and Popular Thought," *Discovering North American Rock Art. L. L.* Loendorf, C. Chippindale, and D. S. Whitley, eds, Tucson: The University of Arizona Press, pp. 15-50.

Loubser, J. H. N. 2007. Recordation, Dating, Analyses, and Management at Paint Rock (31MD379), Appalachian Ranger District, Pisgah National Forest, Madison County, North Carolina, Asheville: unpublished New South Associates report submitted to Pisgah National Forest.

_________. 2013. Graffiti Removal and Camouflage: Writing-on-Stone Provincial Park, Alberta. Lethbridge: unpublished Stratum Unlimited report submitted to Alberta Tourism, Parks, and Recreation.

Loubser, J. H. N. and D. Frink. 2008. Heritage Resource Conservation Plan for Judaculla Rock, State Archaeological Site 31JK3, North Carolina. Sylva: unpublished Stratum report submitted to Jackson County.

Loubser, J. H. N. and R. F. Boszhardt. 2004. "Recordation, Conservation, and Management of Rock Imagery at Samuel's Cave, Wisconsin," *The Rock-Art of Eastern North America: Capturing Images and Insights,* eds, C. Diaz-Granados and J. R. Duncan, Tuscaloosa: The University of Alabama Press, pp. 219-238.

Mooney, J. 1900. "Myths of the Cherokee," *Nineteenth Annual Report of the Bureau of American Ethnology, 1897-98.* Part 1. Washington, D.C.: Government Printing Office.

Parker, G. K. 2006. *Seven Cherokee Myths: Creation, Fire, the Primordial Parents, the Nature of Evil, the Family, Universal Suffering, and Common Obligation,* Jefferson: McFarland and Company, Inc.

Ramsey, J. G. M. 1859. *The Annals of Tennessee to the End of the Eighteenth Century,* Philadelphia: Lippincott.

Raymond, A. 2008. "Approaching Petroglyphs: How Access Affects Management,"

Set in Stone: A Binational Workshop on Petroglyph Management in the United States and Mexico, J. P. Sánchez, A. Sánchez-Clark, and E. L. Abreu, eds, Albuquerque: Petroglyph National Monument, National Park Service, U.S. Department of Interior, pp. 10-19.

Sánchez, J. P., A. Sánchez-Clark, and E. L. Abreu. 2008. eds. *Set in Stone: A Binational Workshop on Petroglyph Management in the United States and Mexico,* Albuquerque: Petroglyph National Monument, National Park Service, U.S. Department of Interior.

Shumate, S, and J. H. N. Loubser. 2011. Phase III Archaeological Investigations at the Judaculla Rock Site (31JK3), Jackson County, North Carolina, Sylva: Unpublished Stratum Unlimited Report Submitted to Jackson County.

Stanley-Price, N. 1994. Introduction to the Management of Rock Art Sites, Anaheim: unpublished paper presented at the 59th Annual Meeting of the Society of American Archaeology.

Trust for Historic Preservation. 2011. *Cultural Heritage Tourism,* Partners in Tourism: Culture and Commerce, *http://www.culturalheritagetourism.org/fourSteps.html*

Wellmann, K. E. 1979. *A Survey of North American Indian Rock Art,* Graz: Akademische Druck und Verlagsanstalt.

Whitley, D. S. 2004. "Management Plan for Rock Art Sites on BLM Lands in California," *The Human Journey and Ancient Life in California's Deserts: Proceedings from the 2001 Millennium Conference,* M. W. Allen and J. Reed, eds, Ridgecrest: Maturango Museum, pp. 225-228.

Wilburn, H. C. 1952. "Judaculla Rock," *Southern Indian Studies* IV. pp. 19-22.

Zeigler, W. G. and B. S. Grosscup. 1883. *The Heart of the Alleghenies,* Raleigh: Alfred Williams and Company.

08

Conservation of the Cultural Heritage and Transformation of the Serrana Society in the Central Highlands of the Baja California Peninsula, Mexico

María de la Luz Gutiérrez Martínez

Introduction

Research and assessment of rock art in Mexico has been slow and complex. Today, rock art is still one of the least studied archaeological materials in the country and, therefore, its importance as cultural heritage is little recognized. This situation is paradoxical considering that there is so much distinctive rock art, especially toward the north of the country, where the geographical and climatic conditions has promoted good preservation and visibility of numerous rock art sites.

Fortunately, the situation is changing. In the 1980s a number of projects were begun that undertook the consistent and systematical study of rock art. The increase in the inventory of rock art sites shows this interest, although it is still insufficient. The Baja California peninsula is a region that exemplifies very well this change of attitude. Located in northwest Mexico, the region remained almost unexplored until very late in the 20th century. Its almost insular status kept native peoples relatively isolated from continental influences, allowing the development of exceptional cultural complexes. Indeed, one of the most outstanding features of the peninsula's prehistory is that these people promoted, in some regions, the mass production of rock art from very ancient times. The inventory of peninsular cave sites currently represents 43 percent of the national total.

The region of the peninsula where this cultural manifestation is most profuse is the central mountain chains. Very often, the landscape is tinted by rock art and their abundance and complexity are sometimes overwhelming. In some regions they are omnipresent; they can be seen in ravines and plateaus, "filos" and "portezuelos," "tinajas" and "malpaisales."[1] The art is integrated into the landscape with remarkable persistence, rubricating it, decorating it and giving it cultural meaning, showing clearly the fluid movement of the people who created them, witnesses of their coming and going (Conkey 1984:264-267; Gutierrez and Hyland 2002:30). In this sense, one of the main values of this region is the landscape itself, understood as the vast space in which the thoughts and memories of its ancient inhabitants were set in stone.

In addition, these mountains were the scene of a rather extraordinary prehistoric event: the development of the rock art tradition of the Great Murals. Harry Crosby coined the term Great Mural (Crosby 1975) to describe these paintings, and it won wide acceptance. The area of distribution of this style includes the mountain ranges of San Borja, San Juan, San Francisco, and Guadalupe. So far the most investigated are San Francisco and Guadalupe, where about 1,150 rock art sites have been recorded, including paintings, engravings, mixed media and geoglyphs (Esquivel 1995; Garcia Uranga 1986; Gutiérrez 1992; Gutiérrez and Garcia 1990; Gutiérrez 2003) (Fig. 1).

Given that the rock art is an exceptional study topic, since 1980 the National Institute of Anthropology and History (INAH) has been carrying out a long-term archaeological research project in the region. In 1997 the first phase of the investigation, which took place at the Sierra de San Francisco and surrounding areas, was completed. This undoubtedly contributed to resolving various issues and priorities of the regional archaeology, and enabled a more focused and productive rock art analysis and laid the groundwork for a future research approach (Gutiérrez and Hyland 2002).

[1] "Filos" are narrow passages that divide the highest parts of streams, i.e., where they are born, which are locally called testeras. "Portezuelos" are low natural passages between ravines. "Tinajas" are natural hollows carved into the bedrock by wind and rain erosion. Because rainfall collects in them, they were essential in determining the seasonal movements of indigenous peoples. These "tinajas" must have embodied deep symbolism for these peoples and in many of them there are concentrations of petroglyphs with terrestrial and aquatic motifs. "Malpaisales" are effusions of lava that are fractured into thousands of blocks. Some of these were ideal for the creation of petroglyphs.

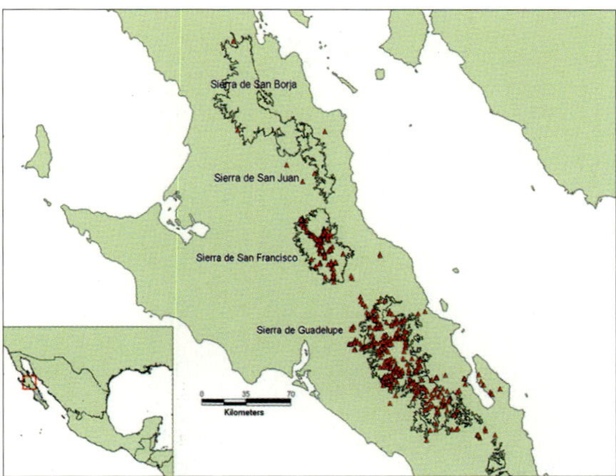

Fig. 1. The Central mountain ranges of the Baja California peninsula. Red triangles indicate the rock art sites located so far.

Sierra de San Francisco

Sierra de San Francisco, in the central highlands, is where the most spectacular and best preserved Great Mural sites are concentrated. It is a small volcanic mountain range located in the northern end of the state of Baja California Sur where high plateaus are sectioned by deep canyons that spread in a radial pattern. Its western slope descends to the vast plains of the Vizcaino desert and the Pacific lagoon systems, while to the east, in the Gulf of California, there are steep mountains (Fig. 2). The mountain range reaches a maximum lift of 1,590 meters above sea level and has an approximate area of 3,600 square kilometers. Its climate is generally dry and warm, as it receives on average less than 100 millimeters of precipitation per year. Thus, surface water sources are few, being rather confined to a few perennial streams and jars. In terms of vegetation, some of the most spectacular communities of the Sonoran Desert are in the area, while relatively lush riparian habitats are found along streams better irrigated in the Sierra. Its main streams are the San Pablo, San Pedro, San Gregorio, San Gregorito, Cuesta Blanca, Los Monos, Santa Martha, Palmarito, El Parral, El Infierno, the Ascension, and El Batequi. These streams and their tributaries contain abundant rock shelters with rock paintings and/or engravings.

Fig. 2. Sierra de San Francisco has high plateaux sectioned by deep canyons. From here a view of its western side; the lagoon systems of the Pacific and the peaks of the Sierra de Santa Clara are quite visible.

Although it is considered one of the most marginal environments on the earth, the Central Desert of the peninsula provided suitable conditions for the establishment of hunter-gatherer-fisher groups, from the terminal Pleistocene (10,000 years BP) to the arrival of Jesuit missionaries at the end of the 17th century. Using the wide variety of coastal, plain, and mountain environments, the Indians followed an intense mobility pattern in search of food, raw materials and water. As a result of this pattern the archaeological sites in the area are many and diverse.

The Pictorial Style of the Great Murals

The rock shelters with Great Mural paintings are the best known of all the prehistoric sites. In terms of scale, this rock tradition is among one of the largest in the world. The images frequently exceed natural size and were created on the high walls and ceilings of the rock shelters, which emphasizes even more their monumentality. Overlapping of figures is very common.

Thanks to the geological nature of this region and the dry weather of the semi-desert peninsula, the rock art is in very good condition, with some large

Fig. 3. East panel of Cueva Pintada, Arroyo de San Pablo, Sierra de San Francisco. Human and animal figures designed in red and black predominate in the Great Mural style.

rock panels having hundreds and even thousands of figures painted in bright colors. The images are essentially realistic in style and are dominated by human figures and land and water animals depicted in red and black, and sporadically in white and yellow (Fig. 3). There are also many sites with petroglyphs, many of which have thousands of individual figures. These are located toward the intermountain plateaus and plains surrounding the mountains. The rock art is just one of the elements that form this extraordinary and ancient landscape. The inscribing of the Sierra de San Francisco on the World Heritage List was based on the outstanding Great Mural sites concentrated here.

The first references to the Great Murals are found in the records of Jesuit priests in the 18th century (Boat 1988). The modern era in research began in the late 19th century when, in 1894 Leon Diguet, an industrial chemist who worked at the copper mine El Boleo, Santa Rosalía, carried out research in the hills of San Francisco and Guadalupe and subsequently published descriptions of several of the sites (Diguet 1895).

The impression of the Jesuits who visited the Great Mural sites was that the paintings were "old."[2] This opinion was not only based on the assessment of the

[2] Reports from the Jesuit Joseph Mariano Rothea and Francisco Escalante are found in Barco (1988:221-212).

physical characteristics of the imagery but more definitively, the responses they got from their informants when asked about the paintings. Local Cochimis groups denied any knowledge of the images and their origins, attributing their authorship to an ancient and extinct race of giants who had come from the North.[3] Given the Jesuit policy of eradicating native religion, the veracity of such responses is open to serious questioning.

Assessment of the Significance of the Sierra

As we have seen, the recognized main value of these mountains is their outstanding rock art, but in order for the significance of these cultural expressions to be preserved, there are other values that need to be kept. The mountains' historical values include prehistoric sites, but also the remains of the missionary period and the survival of the highland cultural traditions whose roots date back to historical events of the 18th century. There are very strong aesthetic values, not only in its spectacular rock art, but also in the beauty of the landscape and vegetation of the canyons and plateaus. The scientific values fall within the scope of the biodiversity and great number of endemic species of flora and fauna, as well as in the study and conservation of rock art sites. Finally, the Sierra has strong social values in terms of the role that culture plays in the conservation of traditional links between mountain communities and the Sud Californianos and Mexicans in general, to contribute to the appreciation of the true story of Baja California from prehistory, through the missionary period up to the present. The "mystery" of the origin of the paintings has long been an important symbolic value; this has dwindled with the advancement of archaeological research, however, for many it will remain an enduring value (Gutiérrez et al. 1996).

The general policy for the management of these resources highlights the preservation of those values that together give meaning to the Sierra, while their conservation should become a source of profit. The development of

[3] Myths tell that gigantic beings were widely reported in Baja California (Barco 1988:209-213) and coincide with the European legends of the Amazons of California. It is interesting to note that the Seri, crossing the Gulf in Sonora also had myths about giants and had the habit of attributing archaeological sites and even recent cultural attributes, to an ancient race of giants (see discussion in Bowen, 1976:103-107).

the area should be sustainable and compatible with the preservation of its educational, historical, and environmental values, thus allowing it to be used and enjoyed by present and future generations.

Cultural Heritage Management

1) Background

The religious evangelization that started on the peninsula at the end of the 17th century led to the complete disappearance of the ancient inhabitants of these lands, the architects of the rock art in the region. The current inhabitants of the Sierra de San Francisco descend from Hispanic mission workers who settled in San Ignacio over 150 years ago or from managers and miners from the French company El Boleo which has extracted copper from the mountains since the end of the 19th century. For years these populations remained almost completely isolated in the interior of the mountains.

Prior to 1970, the mountain residents subsisted by breeding goats and cattle and producing cheese. The population density was low and in general the area lacked roads. These conditions allowed the landscape and the archaeological sites to remain in balance and thus, for years, only the mountain was aware of the existence of these monumental paintings. However, following the expeditions of Erle Stanley Gardner (1962) and Harry Crosby (1997) and the disclosure that was made by them (Hambleton 1979), the situation turned around as the number of visitors to the Sierra increased considerably, creating a mosaic of problems.

Visitors began to hire mountain residents as guides and to rent from them the beasts of burden and mounts necessary for expeditions. This generated an alternative source of income, which undoubtedly triggered major transformations in the lifestyle and thoughts of the local people. In 1990 the mountain residents did not exceed 250, but by 2008, the population had increased to 482 people. This speaks of the demographic changes experienced by the region as a result of the opening of roads and the introduction of new ideas.

2) Santa Martha and San Francisco de la Sierra

Before 1984, tourists that visited these mountains year after year to enjoy the rock art entered the mountains by the only road that existed then: the one that leads to the Valley of Santa Martha, located in the southern foothills of the Sierra de San Francisco. From here they hired guides and rented mounts and beasts of burden that allowed them to reach the canyons where the Great Mural paintings are concentrated. However, in 1984, the construction of the road leading to the town of San Francisco de la Sierra triggered a dramatic change: little by little, the tourists began to prefer the new way because it facilitated access to northern streams, where the most spectacular landscapes and rock art sites are located. Before that, reaching those places required long hours on a mule. With the development of the road, travel time was considerably reduced but the transformation of communities and the environment worsened. An additional aspect that makes this road very appealing is that it gradually ascends to the highest parts of the mountains, allowing visitors to see interesting views of the Vizcaino Desert plain, Pacific lagoon systems, the canyons of the Ascension and St. Paul and the Sierra de Santa Clara, whose solitary summits rise to the east almost bordering the sea (Fig. 2).

Due to the new road, the inhabitants of San Francisco de la Sierra experienced a peak in their activities as providers of tourist services such as guides and carriers, while those in Santa Martha, suffered a decrease in their employment opportunities, even though they were the pioneers in this profession and had the leadership for years. Nowadays, the majority of the guides and carriers are facing a situation of labor shortage and a lack of incentives to improve their activity as guides, which has resulted in poor service and a lack of training of the new generations, who are unfamiliar with the practice and the location of the rock painting sites.

Due to the increase in visitors and a lack of control, some cave sites suffered plundering and informal archaeological excavations. Some paintings were damaged, although fortunately just a little. Visitor undertook more diverse activities, violating established laws when rock art began to be investigated without the knowledge and control of the appropriate authorities. In December 1993 the Sierra was inscribed on the World Heritage list of UNESCO, placing it in the center of international interest.

3) Implementation of the Management Plan

Management of this cultural heritage has been of fundamental interest throughout the years. A strategy was forged gradually in parallel with the archaeological research, and consolidated in 1994 with the consolidation and implementation of the Management Plan of the Sierra of San Francisco. That year the region agreed on two research projects for the mountain rock art, one archaeological and the other about rock art conservation (Stanley 1996).[4] It was then that the conditions were optimal for generating a strategy for protecting the rock art. Several different entities interested in the preservation of rock art agreed on the need to unify criteria and establish a regulatory framework. The Getty Conservation Institute, the Sudcalifornia Friends Association and INAH joined efforts with the aim of designing and implementing the Management Plan.The model adapted for the design of this plan derived from The Burra Charter of Australia ICOMOS (1992)[5] and emphasizes the importance of defining, in the first instance, the significance of this heritage in a way that all policy and management strategies be consistently directed toward the preservation of the values that make it important. Another key feature is the full involvement of all the groups that have an interest in the area under discussion. Note that this participation in the planning process was unprecedented locally.

Thus, representatives of all sectors involved in the cultural heritage of the Sierra including the residents of the mountain communities, ejidatarios, Mexican and foreign tourist service providers, and state and local municipal governments as well as the Secretary of Environment and Natural Resources and the Biosphere Reserve of El Vizcaino. This allowed the reconciliation of interests and decision-making through consensus.

4) The Main Threats

It must be acknowledged that tourism poses the greatest threat to the Great Mural sites. Given the density, distribution pattern and diversity of the prehistoric sites, the region is an extremely exposed archaeological zone, as the sites are everywhere, many of them along the trails that lead to ranches and ravines. Another factor to consider is the vulnerability of the rock art panels, which can suffer irreversible damage by vandalism. In addition, the sites are

[4] Baja California Sur Rupestrian Art Project (INAH, National Archaeological Fund) (1993-1994) and Conservation of Rock Art in Baja California, Mexico (The Getty Conservation Institute) (1994-1995).

[5] See also Pearson and Sullivan (1995)

suffering severe deterioration as a result of the constant visitors, including hikers, as well as the many goats and cattle, which take refuge under the rock shelters. Even the removal of a small amount of soil can alter the archaeological information and there is also the problem of dust deposits that build up on the wall paintings. These are just a few of the factors that are detrimental to the rock art. Therefore, in the beginning, the Management Plan focused on mitigating the impact of visitors on the sites and their surroundings and controlling and monitoring access to them.

Since the 1960s, visitors to the Sierra had established a regular circuit or tour to the most popular Great Mural sites. Thus, one of the immediate priorities was to provide direct protection measures at these sites to reduce damage. These measures included the installation of walkways, railings, fences, access trails, and signs in six of the most visited Great Mural sites. In 2005, an additional site was included: Cuesta Palmarito[6], the most popular site in the southern Sierra (Fig. 4).

5) Policy for Visitor Access

One of the main problems that the area had faced was the uncontrolled access to

Fig. 4. Central Panel of Cuesta Palmarito, Santa Martha brook, Sierra de San Francisco. The picture shows one of the walker breaks.

[6] The inclusion of Cuesta Palmarito was made possible thanks to funding provided by the National Council Adopt a Work of Art AC.

the sites, with or without guides, and consequently, the exploitation of the Great Mural sites from different perspectives. The annual number of visitors to the Sierra had been traditionally low, but with the World Heritage designation the number rose substantially. It was expected that if uncontrolled access continued, the integrity of even some of the more rarely visited sites could be compromised.

The administration of an archaeological zone with hundreds of archaeological sites scattered over thousands of kilometers, needed the design of a sui generis strategy. Certain concepts and guidelines were presented as a preliminary proposal and approved by consensus. These included the extension of the archaeological zone and the rupestrine area; the authorization of access routes to the Sierra; and the designation of different levels of visitor access and restrictions (Gutiérrez and Hyland 2002).

(1) The Archaeological and Rock Paintings Area

Most of the sites with rock paintings and petroglyphs are concentrated in canyons and plateaus. However, their creators extended their territory beyond the foothills of the mountains, toward the Gulf Coast and the Vizcaino Desert plains. Consequently, the archaeological zone surrounds virtually the entire rupestrine area and includes diverse environments such as the Gulf Coast, the slopes of the mountains and the Vizcaino Desert plains, including a connection with San Ignacio lagoon. The archaeological zone is located completely within the Biosphere Reserve of El Vizcaino (Figs. 5, 6).

(2) Access Routes

Two access paths to the mountain were authorized for tourism: one that leads to San Francisco de la Sierra and the other one leading to Santa Martha Valley. Both villages—San Francisco de la Sierra and Santa Martha—are the only allowed departure points for entering to the rock art area. Expeditions are prepared here as it is where guides and wranglers can be hired, and mounts and beasts of burden can be rented (Fig. 7).

(3) Levels of Visits

Visits to the Sierra were classified into four levels. Level I includes the sites easily accessible by vehicle and limited hiking; Level II consists of selected sites within San Pablo Creek and the Arroyo del Parral that are accessible by mule or hiking from the Santa Martha and San Francisco de la Sierra villages and requires camping. Camping is allowed only at designated sites. In order to avoid environmental degradation of the areas and streams and to prevent

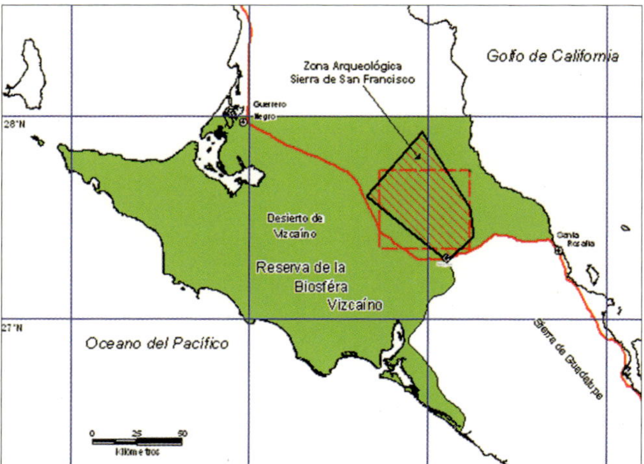

Fig. 5. Fortunately the Sierra de San Francisco is located entirely within the Biosphere Reserve El Vizcaino, which contributes in the protection of its cultural resources.

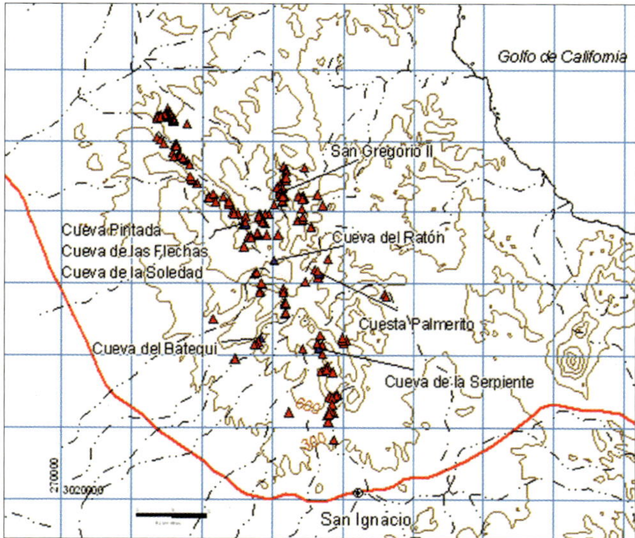

Fig. 6. Great Mural sites in the Sierra de San Francisco. Some of the most popular are highlighted.

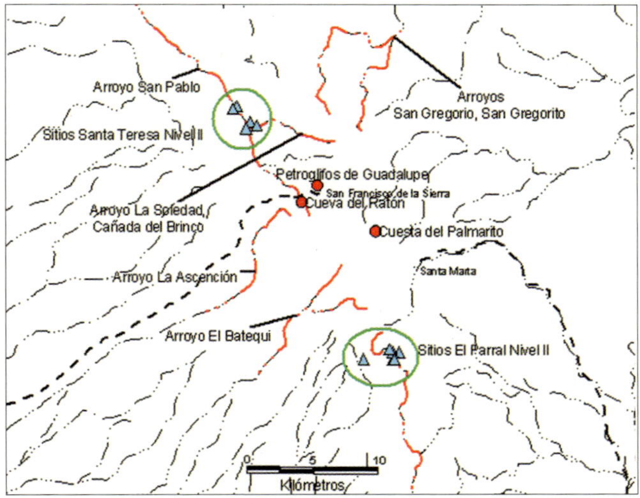

Fig. 7. Authorized access routes and levels of visit

saturation of the Great Mural sites, a maximum number of visitors has been defined for both sites and camps. Level III includes other sites that are in less frequented areas such as the Arroyo de San Gregorio, San Gregorito and of the Batequi, which can be visited only by filing a permit two weeks in advance. Authorized groups are accompanied by a custodian of the INAH. Level IV is designed for purposes of research and access is only permitted to investigators accredited by the INAH and the Biosphere Reserve El Vizcaino (Fig. 7).

This system allows visitors to experience a wide range of sites and at the same time protects the majority of the sites that are fairly well preserved. In this sense the most popular sites have remained open under this Management Plan.

(4) Surveillance

As a specific issue of tourism administration, a process of monitoring has been established. Based on an observation of the number of visitors who come to the mountains throughout the year, their preferred routes and interests have been determined. Monitoring involves the inspection of site conditions and their environment, as well as the implementation of regulations concerning visitors and guides. They include the following actions: regular trips by custodians over established routes which cover the whole of the rock art area; increased surveillance in all areas during the high tourist season (from October to May); random inspections of expeditions in areas open to the public; and, participation of the custodians in monitoring authorized expeditions to restricted areas.

In this sense, a crucial measure was the establishment of an Information Module in San Ignacio, the nearest town to the mountains. Through this, the presence of the INAH in the region is guaranteed, while it plays the dual role for the public of being an interpretation center and a reservations and guidance center for visiting the Sierra.

Fundamental to the successful conservation of the paintings is the active participation of the local population. Adequate protection of sites depends on local guides who accompany visitors, and three INAH custodians, residents of the Sierra, who deal with monitoring the north, center and south sectors of the Sierra.

(5) Evaluation of the Plan

To ensure the success of the Management Plan, monitoring and periodic review are essential. Every two or three years an evaluation meeting is held in San Ignacio in which an assessment of the results is made, as well as a review of the problems that have been detected, to first and foremost find mutually acceptable solutions. There have been four evaluation meetings so far.

One of the cornerstones to the success of this type of management plan and, let's say, the spirit that drives it and makes it valid, is having a consensus. This implies that no decision will be taken unilaterally and outside of evaluation meetings. We must remember that this is a flexible strategy that can evolve according to the development needs in the region. However, there are core aspects that cannot be changed, I mean those that are directly related to the stability of rupestrine art. If they were to change, it would impact directly on the rock art by causing deterioration and, then, the Management Plan would lose its raison d'être.

Significant progress has been made over the years regarding the management and protection of this cultural heritage. However, there have been events that have infringed the Management Plan, and now it is time to balance the current status of this management.

Recent Problems

In 2002 a series of events were triggered that threatened to destabilize the strategy. These events began with the Government's implementation of regional development programs with state, national and international cooperation that were proposed to the Sierra communities.[7] While the actions arising from these projects posed very laudable objectives, those responsible ignored the background around the management of the cultural property which INAH had developed in the region since 1980 and, above all, it was clear they were absolutely ignorant of the idiosyncrasies of the mountain inhabitants and visitors and the vulnerability of the rock art. INAH was the subject of a smear campaign by those responsible for these programs and it negatively impacted the people who till then had been our allies in the protection of the heritage—the mountain guides.

By political lobbying, the director of INAH in Baja California Sur took the unilateral decision to authorize the entry of visitors to the mountains without any registration. This decision showed at that time, two very serious aspects: 1) passed over the spirit of the Management Plan, which claimed that any change must be approved by consensus, and 2) left to set a precedent of high risk to the sustainability of the Plan as any other concern by the community, the Government of the State or public or private organizations, through the community, would move to the political arena. Since then, I pointed out the danger that through coercion, one by one, all the measures implemented for the preservation of heritage would collapse, and unfortunately I was right.

In 2006 the Government of the State of Baja California Sur unveiled a program for the opening of a network of trails in the Sierra de San Francisco. Coincidentally, this program emerged parallel to the regional development plan named Sea of Cortez, which was designed under the shadow of very powerful business and political interests. Under this type of "Development Plan," the hopes of the communities are manipulated, high expectations are generated and the cultural object is transformed into a lucrative political booty that can be exploited without taking into consideration its vulnerability and values involved and should be protected.

[7] Integrated Project of Sustainable Development of the Reserve of the Biosphere Vizcaino. International Cooperation Spanish Agency. Araucaria Project.

Given the expectations generated by the project, it is not difficult to conclude that what was sought with these roads was to generate the essential infrastructure to transport tourists in vehicles to the canyons of the Sierra. If built, many of these roads would reach the core of the rupestrine area where the cultural landscape is primordial and rock shelters contains an exceptional Great Mural panels. If this "input output" tourism is allowed to enter the areas where at the present time you can visit through eco-expeditions which last for three to four days, most of the guides will lose the ability to make rational use of part of the heritage that so far has been the object of cultural tourism.

After facing pressures and very unpleasant situations with very low profile political actors, and thanks to the timely intervention of the authorities of the Biosphere Reserve El Vizcaino, the opening of roads project was stopped, although there is a chance the project will be reactivated sooner or later. We must be very alert whenever election periods arise, because it is a fact that political events like elections trigger or accentuate these problems. It was not possible to stop the paving of 30 of the 37 kilometers of the road which leads to San Francisco de la Sierra. In a few more months, there will be a fully paved road and it is quite possible that this "facility" will exert new pressures on the Great Mural sites.

We all know what a road means for an area like this; the risks are imminent. Providing the Sierra with essential infrastructure could, in the not-too-distant future, open the door to urban development that would impact those values that still make this an exceptional region. Fig. 8 shows clearly how

Fig. 8. Road to San Francisco de la Sierra. The sign reads: "The Federal Government modernizes a stretch of 9.5 kilometers benefiting 9,000 people to better living." We know that the population of the Sierra does not exceed 500 people.

the Government manipulates information for political purposes and although I am not sure, I think that new scenarios are being prepared: here it is indicated that this road will enable "9,000 people to live better." This is also false and absurd. We know from the most recent census that the total population of the area is 500.

Conclusion

What are the strengths of the Management Plan 20 years after its implementation? The management model has proven to be an effective strategy that has contributed to the preservation of the historical, aesthetic, scientific, and symbolic values of the region. The landscape, archaeological and historical sites, petroglyphs and cave paintings have remained stable. Archaeological research was regulated and there has been no looting of archaeological items as before.

What are the weaknesses? The Management Plan has proved ineffective to assist in the preservation of the social values of the region, for lack of space it is not possible expound how the whole process has been, but I must say that there has been a gradual loss of traditions and cultural identity. While the number of visitors to the mountain range has remained relatively stable, the number of guides has doubled in some places while job opportunities have decreased in others causing communities to confront each other. Finally, it is necessary to point out that the Management Plan does not have a legal framework that will provide support and strength and it is clear that it is weak against the impending onslaught of "development and modernity."

What lessons have we learned over the years? During the process of design and implementation of the Management Plan, we knew the importance of reconciling interests, making decisions through consensus and, above all, involving representatives of all institutions which at that time were related to the Sierra. However, we did not anticipate that over the years, the environment, people and interests would change. To mention some examples: a) when the present Management Plan was signed, the constitutional amendments that led to the alienation of ejido lands had not been approved[8] turning the ejidatarios of Baja California Sur into large landowners, and now the best land has been taken over by businessmen and politicians; b) the main limitation to large populations and tourism development in the region was the scarcity of water, but today there are desalinization plants producing high quality drinking water, and this has raised the land prices throughout the region; c) the Government of the State of Baja California Sur transformed its management style when there was a switch in political parties, with excessive populism being a main feature of the new administration. All this has impacted the people and their interests, causing irreversible transformation of the mountain culture, one of the most important values that were unique to the Sierra de San Francisco and even the region. This is a great lesson: when there are marginalized population centers in "protected areas," it is very difficult to keep the scenarios in which a management model was generated. Controlling the clamor for "development" and the political use of cultural heritage is hard work and sometimes almost impossible to achieve.

The abundance of rock art sites located in the central mountain ranges of the Baja California peninsula and the enormous surrounding area and the scarcity of human and financial resources makes it extremely difficult for us as administrators to protect the mega archaeological region. Experience has shown that without the support of the local population and without the understanding and support of the three sectors of government, any attempt to preserve the natural and cultural values of this region will fail. It is therefore essential that the private and public sectors increase their efforts to reach agreements that will allow us to establish lasting ties of friendship and strategic alliances with the mountain communities who have lived with this heritage for generations, and have been the first line of defense that we have.

Despite some setbacks, we will continue emphasizing both the importance of preserving this magnificent rock art as well as the importance that interdisciplinary research has to achieve its proper management. Let us remember that heritage acquires a social value from the moment a group of people recognize it as such and it is important for their identity, and that is when archaeology transcends the communities coexisting with that heritage, not only because it provides elements to give cohesion as a society, but also because it can contribute to their sustainable development.

[8] Certification Program of Ejido Rights (PROCEDE)

BIBLIOGRAPHY

Barco, Miguel. 1988. *Natural History and Chronicle of Old California.* Universidad Nacional Autonoma de Mexico, Mexico D.F.

Bowen, T. 1976. "Seri Prehistory: The Archaeology of the Central Coast of Sonora, Mexico," *Anthropological Papers of the University of Arizona,* 27.

Conkey, Margaret W. 1984. "To find Ourselves: Art and Social Geography of Prehistoric Hunter-gatherers," *Past and Present in Hunter-Gatherer Studies,* edited by Carmel Schrire, Academic Press, New York, pp. 253, 276.

Crosby, Harry W. 1997. *The Cave Paintings of Baja California: The Great Murals of an Unknown People,* Copley Books, Salt Lake City, USA.

Diguet, L. 1895. "Note sur Pictographie de la Basse-Californie," *L'Anthropologie* 6, pp. 160-175.

Esquivel, L. 1995. *Report of the Sierra de Guadalupe Archaeological Project,* Field seasons 1992/1993, Delivered to the Archaeology Council of the INAH, Technical file, Mexico, D.F.

Garcia-Uranga, Baudelina. 1986. *Report of the Work Carried out during the First Field Season of the Project Location, Record and Study Sites with Rock Paintings and/or Petroglyphs on the Peninsula of Baja California, Mexico,* Delivered to the Archaeology Council of the INAH. Technical file, Mexico, D.F.

Gardner, E. S. 1962. *The Hidden Heart of Low,* Wm, Morrow, New York.

Gutiérrez, M. María de la Luz. 1991. *Report of the Work Carried out during the Second Season Field of the Location, Record and Study Sites with Rock Paintings and/or Petroglyphs Project in the Sierra de San Francisco, B.C.S.* Delivered to the Archaeology Council of the INAH, Technical file, Mexico, D.F.

_________. 2003. *Second Annual Technical Report of the Social Identity, Ritual Communication and Rupestrian Art Project: The Great Mural of the Sierra de Guadalupe, B.C.S.* Delivered to the Archaeology Council of the INAH, Technical file, Mexico, D.F.

Gutiérrez, M. María de la Luz and Baudelina Garcia-Uranga. 1990. *Contextual Analysis of Cave Painting: A Study Case in Baja California,* Bachelor thesis in archaeology without publishing, National School of Anthropology and

History, SEP - INAH.

Gutiérrez, M. María de la Luz and Justin R. Hyland. 2002. *Archaeology of the Sierra de San Francisco: Two Decades of the Great Wall Phenomenon Investigation,* Scientific collection, no. 433 1st. Ed, National Institute of Anthropology and History, Mexico, D.F.

Gutiérrez, M. María de la Luz, Justin R. Hyland, Enrique Hambleton, and Nicholas Stanley P. 1996. "The Management of Word Heritage Sites in Remote Areas," *Conservation and Management of Archaeological Sites* in 1(4), James & James. London, pp. 209-225.

Hambleton, Enrique. 1979. *Rupestrian Paintings of de Baja California,* Fomento Cultural Banamex, Mexico D.F.

Marquis-Kyle, P. and M. Walker. 1992. *The Illustrated Burra Charter, ICOMOS, Australia.*

Pearson, M. and Sharon Sullivan. 1995. *Looking after Heritage Places: The Basics of Heritage Planning for Managers, Landowners and Administrators,* Melbourne.

09

Conservation and Management of Australia's Kakadu National Park and its Petroglyphs

Jiyeon Kim

Kakadu National Park, covering almost 20,000 km^2 in Australia's Northern Territory, contains about 5000 Aboriginal petroglyph sites that are recognized as vital cultural assets. As such, the park was inscribed on UNESCO's World Heritage List in 1981 for both its cultural and natural values, becoming one of the first sites in Australia approved for inscription. Notably, the park's designation as a World Heritage Site emphasizes the natural environment and ecological system of the area as key components of its heritage value. As such, the petroglyphs and other historic sites are recognized as integral elements of the park's complex and ever-changing value system, rather than simply as isolated historical evidence or objects of appreciation.

Having first been designated as a national park and then inscribed on the World Heritage List, the site has attracted much public attention and has been the subject of much political debate. The park is now jointly managed by the Australian government and the Aboriginal people who still live in the area and make their living from the region's natural and cultural resources. The system of joint management has been difficult at times, as settler Australians and Aboriginal landowners often have different priorities and values, different ideas about preservation, and different concepts of nature and culture. Debates over these issues have often been exacerbated by political conflicts between the governments of the Northern Territory and the Commonwealth of Australia.

In this chapter, I will first examine the petroglyphs in the Kakadu National Park area, which were produced over an extremely long time span. These petroglyph sites have long been preserved, managed, and shared as a vital part of Aboriginal history, and the methods and principles of their preservation have constantly evolved, even since the establishment of the national park. To better understand the current direction of the management, I will also discuss the history of Kakadu National Park's establishment and administration. My hope is that this case can serve as an example of how cultural values were defined and re-defined, and how different groups with conflicting interests were able to reach agreement.

History of Aboriginal Occupation and Colonization

The territory of Kakadu National Park features tremendous biodiversity, the result of several periods of abrupt climate change in the area's history. Until about 13,000 years ago, the land was very dry, but the rise of global temperatures and the subsequent melting of ice produced more humidity, favorable for the growth of vegetation. Beginning around 8,000 years ago, rainfall increased, leading to flooding of river valleys and the formation of forests and salty mudflats. Finally, between 4,000 and 1,500 years ago, the salty estuarine conditions were replaced by freshwater swamps (Press and Lawrence 2007:18).

The area contains many archaeological sites, including the petroglyph sites. The dating of these sites varies, but it is generally considered that the area has been inhabited for more than 50,000 years, and some of the rock paintings are more than 20,000 years old (Ovington 1986). The early inhabitants who produced the petroglyphs are thought to have been hunters (Press and Lawrence 2007:19-20), and they had quite advanced technology, including various types of stone tools. When the sea level rose, people moved inland and the population considerably increased. At that time, they began using stone tools with sharp points and making tools from bone, wood, and shell (Ibid).

Until recently, Aboriginal people still used the land for hunting and fishing, but over the last few centuries, the social and economic landscape of the region has changed drastically with the arrival of various non-Aboriginal people.

Nonetheless, for the most part, the region remained pastoral, with most jobs related to livestock and a small-scale mining industry. In the late nineteenth and early twentieth century, missionary work began, and buffalo hunting emerged as a major seasonal industry that employed many Aboriginal people (Lawrence 2000:24). The men typically worked as guides and other staff in the buffalo camps, while the women processed the hides and did other chores. The tourism industry remained in its infancy until the 1970s, but gradually developed through the latter half of the twentieth century (Lawrence 2000:27). But the most striking change since the arrival of the settlers was the rapid decline in the Aboriginal population. The main cause of the decline was the spread of introduced diseases, but local people also sometimes refer to violence among the Aboriginal people themselves (Levitus 2007:86). In response to the social disruption, the administration of the Northern Territory established a reserve for the Aborigines, which became part of the Aboriginal Freehold Land in the 1970s, and then later part of the Kakadu National Park. With the establishment of the national park, ownership of the land was granted to Aboriginals, leading to a significant expansion of Aboriginal residence in the park area (Levitus 2007:91-92).

Petroglyphs of Kakadu National Park and Their Preservation

Kakadu National Park contains about 5,000 petroglyph sites, mostly paintings. The petroglyphs were the first feature of the area to draw international attention, and were a major reason for the inscription of the park on the UNESCO World Heritage List in 1981.

Rock art has long been a crucial part of the lives of the Australian Aborigines. According to some beliefs, the Creation Ancestors manifested themselves in rock paintings, and so the sites became *djang* (dreaming places), sacred areas that could only be seen by elders. Perhaps most importantly, these rock paintings remain highly relevant to today's Aboriginal people. The paintings were produced over an extremely long period of time, from ancient times through the arrival of the settlers. They are essential artifacts that not only reflect the rituals and sorcery of ancient times, but also document the animals and tools introduced through contact with the Europeans.

The rock paintings can be categorized into several types, based on representational method. One of the first paintings was made by pressing paint-covered hands, grass, or other objects to the rock walls. Later, hand prints were made by blowing a pigment solution from the mouth over hands pressed on rock (Ovington 1986:32) (Fig. 1). Other old styles include large naturalist paintings of people and animals, drawn first in detailed outline and then filled in with an ochre wash. Also of note are the figures drawn with simple lines (Fig. 2), images of people, animals, and animal-headed beings, which are thought to be related to hunting (Ovington 1986:33). Petroglyphs of the later periods are marked by the appearance of "X-ray" paintings (Fig. 3), which displayed the internal organs and bone structures of animals (Ovington 1986:36). With

Fig. 1. Hand Stencils, Kakadu National Park, Australia (Photo: Shutterstock)

Fig. 2. Spirit, Kakadu National Park, Australia (Photo: Shutterstock)

Fig. 3. Fish, Kakadu National Park Australia (Photo: Shutterstock)

Fig. 4. Namandi, Kakadu National Park, Australia (Photo: Shutterstock)

the incursion of freshwater and the arrival of the settlers, new species, such as magpie geese, began to be depicted. The X-ray style continued, but rather than indicating the inner organs, the patterns became more decorative (Fig. 4). Non-aboriginal people, primarily Macassans and Europeans, were also depicted, along with their ships, horses, and weapons.

The petroglyph sites and other archeological sites of the park have received international attention from early on, resulting in relatively strong measures for surveillance and preservation. Public access is provided to three major petroglyph sites at Burrunggui, Ubirr, and Naguluwurr. According to the park management plan, "access to public rock art and archaeological sites needs to be managed carefully and information regarding the conservation requirements of these sites should be available for park visitors" (National Parks, Australia 2007:51). Peter Wellings, former manager of Kakadu National Park, summarizes the main management objectives related to petroglyph sites as follows:

- To recognize the interests of site custodians and to consult with them regarding all aspects of site management;
- To preserve the sites;
- To systematically record all sites in the park and to establish a program for their ongoing management and conservation;
- To interpret Aboriginal prehistory and rock art to visitors by developing selected sites for visitors viewing and providing information on Aboriginal culture and site (Wellings 1995:251).

For detailed conservation measures, the management plan also states that "serious physical damage to rock art may be caused by water flowing over the rock and removing pigment or depositing salts and minerals on the painted surface. Other damaging agents include vegetation, fire, mud-building wasps, termites and people touching or vandalising paintings. Feral animals may also damage archaeological sites" (National Parks, Australia 2007:51). The following activities are summarized as being of top priority:

- Site recording using photogrammetry and other documentation techniques: Once sites are located, site recording and survey photography, including photogrammetric recording, takes place to establish a base record from which physical changes can be measured.

- Site protection: This includes installation of silicone drip lines where appropriate to divert water from painted surfaces, removal of wasp nests and removal of vegetation which presents a threat to painted surfaces. Where appropriate, fences may be constructed to keep visitors and animals, such as feral pigs, from touching painted surfaces.
- Research into factors contributing to the deterioration of painted surfaces and techniques for the restoration of damaged art.
- Development of infrastructure: Where a decision is taken to open a site for public use, it is usual to install appropriate infrastructure to accommodate tourist traffic, e.g. installation of defined walking tracks, boardwalks, and stone paving to direct visitor traffic and to reduce dust levels.
- Development of appropriate signage and information for tour operators and park visitors that engenders respect for the art and awareness of its importance and threats to its conservation.
- Supervision of visitors at art sites (Wellings 1995:251-252).

While these sites are being preserved according to international standards, more sensitive issues emerge in those areas still inhabited by Aboriginal people, particularly with regards to sites that have retained spiritual meaning. Wellings points out that project management should be based on the employment of Aboriginal people to carry out site recording and maintenance work (Ibid). The recent management plan also emphasizes the primary importance of consulting Aboriginal representatives in all matters related to the research, management, and preservation of the park's cultural and natural heritage, including tourism (National Parks, Australia 2007:51). Therefore, the management and conservation of the petroglyphs and other historic sites of Kakadu Park cannot be separated from the larger discussion of how to define and interpret the park's cultural heritage and its value. This discussion dates back at least to the 1970s, when the area was first designated as a national park, and became more heated over the next two decades, as the park was inscribed as a UNESCO World Heritage Site.

Establishment of the Park and Inscription on the UNESCO World Heritage List

In the 1970s, large deposits of uranium were discovered in the Alligator Rivers Region, precipitating the settlement of the ongoing discussion about the Aboriginal ownership of the reserve land and the establishment of a national park in the area. Mr. Justice Woodward, who was commissioned to solve the land dispute, suggested the establishment of a national park in the region, to be jointly managed by Aboriginal people and the commonwealth "in the context of reconciling Aboriginal interests with conservation" (Press and Lawrence 2007:2). Woodward prepared a list of key principles for guiding the joint management of the park, and these principles remain the basis for the park's management to this day. The principles basically give precedence to the protection of Aboriginal rights in matters related to conservation. For example, the principles state that "Aboriginal people should not be outvoted by conservation interests without having their point of view considered by individual adjudicator," and that "attempts should be made to reconcile Aboriginal interests and the interests of conservation by compromise, even though the results may not be best in terms of conservation planning" (Aboriginal Land Rights Commission 1974, quoted in Lawrence 2000:83). The last principle is that "Aboriginal interest should be over-ruled only in cases where the conservation case is strong: for example, in the preservation of endangered or vulnerable species" (Ibid).

In 1976, the Commonwealth of Australia set up the Ranger Uranium Environmental Inquiry, to "reconcile land use matters and minimize any potentially adverse social, cultural, and biological consequences of mining and tourism" (Ovington 1986:1). The commissioner recommended that Aborigines be granted land rights for part of the Alligator Rivers Region and that a major national park be established under Commonwealth legislation (Ibid). Stage 1 of Kakadu National Park was proclaimed as a National Park in 1979, followed by stages 2 and 3 in the 1980s. Soon after the designation of the national park, Kakadu was inscribed on the UNESCO World Heritage List, with stage 1 being inscribed in 1981 and the entire park being inscribed in 1992. The park was inscribed for both cultural and natural values, under the criteria (i), (vi), (vii), (ix), and (x), with details as follows:

Criterion (i): Kakadu's art sites represent a unique artistic achievement because of the wide range of styles used, the large number and density of sites and the delicate and detailed depiction of a wide range of human figures and identifiable animal species, including animals long-extinct.
Criterion (vi): The rock art and archaeological record is an exceptional source of evidence for social and ritual activities associated with hunting and gathering traditions of Aboriginal people from the Pleistocene era until the present day.
Criterion (vii): Kakadu National Park contains a remarkable contrast between the internationally recognised Ramsar-listed wetlands and the spectacular rocky escarpment and its outliers. The vast expanse of wetlands to the north of the park extends over tens of kilometres and provides habitat for millions of waterbirds. The escarpment consists of vertical and stepped cliff faces up to 330 metres high and extends in a jagged and unbroken line for hundreds of kilometres. The plateau areas behind the escarpment are inaccessible by vehicle and contain large areas with no human infrastructure and limited public access. The views from the plateau are breathtaking.
Criterion (ix): The property incorporates significant elements of four major river systems of tropical Australia. Kakadu's ancient escarpment and stone country span more than two billion years of geological history, whereas the floodplains are recent, dynamic environments, shaped by changing sea levels and big floods every wet season. These floodplains illustrate the ecological and geomorphological effects that have accompanied Holocene climate change and sea level rise.

The Kakadu region has had relatively little impact from European settlement, in comparison with much of the Australian continent. With extensive and relatively unmodified natural vegetation and largely intact faunal composition, the park provides a unique opportunity to investigate large-scale evolutionary processes in a relatively intact landscape.

Kakadu's indigenous communities and their myriad rock art and archaeological sites represent an outstanding example of humankind's interaction with the natural environment.
Criterion (x): The park is unique in protecting almost the entire catchment of a large tropical river and has one of the widest ranges of habitats and greatest number of species documented of any comparable area in tropical northern Australia. Kakadu's large size, diversity of habitats and limited impact from

> European settlement has resulted in the protection and conservation of many significant habitats and species. The property protects an extraordinary number of plant and animal species including over one third of Australia's bird species, one quarter of Australia's land mammals and an exceptionally high number of reptile, frog and fish species. Huge concentrations of waterbirds make seasonal use of the park's extensive coastal floodplains (UNESCO 2013).

The omission of criterion (iii) is noteworthy. The current definition of criterion (iii) is "to bear a unique or at least exceptional testimony to a cultural tradition or to a civilization which is living or which has disappeared"; however, in 1991, the definition did not account for living civilizations. Thus, as peculiar as it may sound, the fact that the people who had created the cultural tradition were still living there precluded the park from meeting this criterion. Despite this anomaly, most contemporary accounts of the site's cultural values strongly emphasize the continuity of the Aboriginal cultural heritage. According to the 1991 nomination of the entire park area as a World Heritage Site, the "cultural sites of Kakadu National Park are of great antiquity and exhibit great diversity both in space and through time, yet the overwhelming picture is one of continuous cultural development; the sites represent a unique artistic achievement that is part of a living tradition" (quoted in Lawrence 2000:219). The same points are echoed in the current summary of the site's Outstanding Universal Value:

> The hunting-and-gathering tradition demonstrated in the art and archaeological record is a living anthropological tradition that continues today, which is rare for hunting-and-gathering societies worldwide. Many of the art and archaeological sites of the park are thousands of years old, showing a continuous temporal span of the hunting-and-gathering tradition from the Pleistocene Era until the present. While these sites exhibit great diversity, both in space and through time, the overwhelming picture is also one of a continuous cultural development (UNESCO website 2013).

In sum, the Aboriginal cultural heritage is considered to be still very much alive and evolving. One term that frequently appears in this context is "cultural landscape." According to the 1995 World Heritage Convention, cultural landscape refers to social and cultural change that occurs over time under the constraints imposed by the natural environment (quoted in Palmer 2007:261).

This concept is well defined in the Visitor's Guide to the park, which states that Kakadu's status as a cultural landscape means that it has been shaped by the history, language, kinship, and ecological knowledge of the Aboriginal ancestors (National Parks 2007:261). However, the cultural landscape of Kakadu also encompasses continuous interaction between humans and nature, which includes interactions between different people in terms of their ways of utilizing nature.

Reconciling Conflicting Values

Any attempt to discuss continuity or define Kakadu's cultural landscape entails consideration of the competing interests between Aborigines and settler Australians. For the Australian federal government, the park is primarily a place for conservation, a natural and cultural heritage site valued for its past and timeless beauty. For Aboriginal landowners, on the other hand, Kakadu is their homeland, the site of their everyday activities and production. The region has a long history of human settlement, and the question of how that history is incorporated into its "heritage value" must be raised. Notably, discussions of Aboriginal land rights and the idea of designating the area as a national park did not begin in earnest until 1970, with the discovery of uranium and the subsequent establishment of the uranium mining industry (Press and Lawrence 1995:2-3). Kakadu National Park is now heralded as a bastion of wilderness, where pristine nature is preserved, but it must be acknowledged that "nature," at least in the context of cultural heritage, is socially and historically constituted (Palmer 2007:256).

The designation of Kakadu National Park as a World Heritage Site aroused conflict between the Commonwealth of Australia and the Northern Territory government, but notably, the Aboriginal landowners were not directly involved in the dispute. The conflict between the two governments has a long political history behind it, but it was also related to the question of development versus. conservation. Basically, the Northern Territory believed that the World Heritage status would impede the industrial and economic activities of both the Aborigines and the settlers. As mentioned, uranium mining was an essential factor that sparked the creation of the National Park, and it remains an ongoing concern for park management and UNESCO, even after the granting of World Heritage status.

In the 1980s, one of the main points of contention between the two sides was the validity of the Commonwealth's assessment of the area's cultural heritage. A Conservation Commission hired by the Northern Territory severely criticized the Commonwealth's assessment, claiming that it did not reflect the Aboriginal view. The committee stated that "Aboriginal values tend to become submerged in western interpretation. There is no certainty that objects which are recognized by western eyes as culture or art have universal value" (Conservation Commission of Northern Territory 1987, quoted by Lawrence 2000:230). This evaluation is connected to more recent criticisms against environmental purists, who wish to see Kakadu kept as a timeless reserve where nature can be aesthetically enjoyed and explored in its primeval form, rather than as a vital part of contemporary society and life (Palmer 2007).

For the last few decades in Australia, Kakadu National Park has been at the center of public debates about development, preservation, Aboriginal land rights, and the extent of the Aborigines' control over their own cultural heritage. As such, the park has at times been used as a pawn in the power game between local and federal governments. As a World Heritage Site with recognized cultural value, the park is considered to be an outstanding example of "people's interaction with their natural environment" (Australian Government website 2013). In the 1980s, when Kakadu National Park was first inscribed on the World Heritage List, its cultural heritage was rather narrowly defined, but since then its meaning has broadened, most representatively in conjunction with the concept of cultural landscape. In the same vein, the petroglyph sites of Kakadu have begun to be understood as an essential part of not only the past, but also the present Aboriginal life. The rock art is still very relevant to the life and belief of the landowners; it must be, and is being preserved and maintained with this understanding. Decades of debates on the cultural values of Kakadu National Park have yielded some agreement between opposing parties, but new issues and concerns continue to arise, particularly related to the area's rich uranium deposits, which are drawing increasing interest from global corporations. The case of Kakadu, however, serves as an excellent example of how seemingly divergent interests and values can be reconciled through ongoing discussions and negotiations.

BIBLIOGRAPHY

Books and Articles

Lawrence, David. 2000. *The Making of a National Park,* Melbourne University: Melbourne, Australia.

Levitus, Robert. 1995. "Social History since Colonisation," Presss T. D. Lea, A. Webb, and G. Alistair eds. *Kakadu: Natural and Cultural Heritage and Management,* The Australian National University: Darwin, Australia.

Ovington, Derrick. 1986. *Kakadu: A World Heritage of Unsurpassed Beauty,* Canberra: Australian Government Publishing Service.

Palmer, Lisa. 2007. "Interpreting 'Nature': The Politics of Engaging with Kakadu as an Aboriginal Place," *Cultural Geographies* 14.

Press, Tony and David Lawrence. 1995. "Kakadu National Park: Reconciling Competing Interests," Presss T. D. Lea, A. Webb, and G. Alistair eds. *Kakadu: Natural and Cultural Heritage and Management, Australian Nature Conservation Agency,* North Australia Research Unit, The Australian National University: Darwin, Australia.

Wellings Peter. 1995. "Management Consideration," Presss T. D. Lea, A. Webb, and G. Alistair eds. *Kakadu: Natural and Cultural Heritage and Management,* The Australian National University: Darwin, Australia.

Official Document

National Parks, Australia. 2007. "Kakadu National Park Management Plan 2007-2014."

Official Websites

Department of Education, Australian Government. 2013. "Kakadu National Park," *http://www.environment.gov.au/topics/national-parks/kakadu-national-park*

Kakadu National Park Official Website. 2013. *http://kakadu.com.au/*

UNESCO World Heritage List. 2013. "Kakadu National Park," *http://whc.unesco.org/en/list/147*

10

Thin-line Engravings and Inscriptions of the Cheonjeon-ri Petroglyphs in Ulsan

Hotae Jeon

Location, Contents, and Current Research

1) Location

The Cheonjeon-ri-seoseok ("lettered stone of Cheonjeon-ri," i.e. the Cheonjeon-ri petroglyphs, Cheonjeon-ri Gakseok) is located at San 210, Cheonjeon-ri, Dudong-myeon, Ulju-gun, Ulsan, around the mid- to upper reaches of Daegok Stream, which also flows in front of the Bangudae petroglyphs in Daegok-ri, Ulju. The Bangudae petroglyphs are about two kilometers away, and can be found by following the steep slopes of a high rock wall across from Cheonjeon-ri. The Cheonjeon-ri petroglyphs are carved on a large, screen-like rectangular panel of rock under the ridgeline, as well as on four other rock panels north of the main panel (Fig. 1). The rock panels are shaded all day, except very briefly at midday, because the sun is blocked by a mountain on the other side of the stream. Along the stream in front of the site are several broad rock beds, spacious enough to hold more than ten people. These rock platforms also contain more than 100 dinosaur footprints.

2) Contents

The Cheonjeon-ri petroglyphs, which have now been designated and preserved as Korean National Treasure 147, were discovered on December 25, 1970, by a

Fig. 1. East view of the Cheonjeon-ri Petroglyphs

Fig. 2. East view of main rock panel of the Cheonjeon-ri Petroglyphs

group of researchers from Donguk University who specialized in Buddhist art (Mun 1973; Hwang and Mun 1984).

The large rock wall where the petroglyphs were carved is 2.7 meter high and 9.5 meter wide (Fig. 2). Other related rock panels range from about 1.0 to 2.5 meter in size. The large rock panel features carvings of dogs, deer, dragons and other imaginary creatures, various human figures, a procession of people, ships, and many different geometric patterns (Fig. 3). In addition, there around 1,000 inscriptions that were added during the Silla period (Hwang and Mun 1984). The carvings were rendered with sharp metallic tools, using different techniques, including silhouette chipping, deep outline chipping, and thin outline incision. Most of the animals and people are represented with silhouette chipping, while deep outline chipping was used for the geometric patterns, and the procession, ships, and imaginary creatures were done mostly with incision. The following describes the distribution of the petroglyphs, categorized by carving technique.

(A) The center section of the main rock panel contains many geometric patterns and inscriptions, while most of the figurative images are found to the right and left. For a more detailed description, the panel can be vertically divided into three sections (i.e. left, center, right), and each of these sections can be subdivided into upper and lower parts.

① Within the left section of the main rock panel, the upper left and central areas are covered with figures chipped in silhouette, including numerous pairs of deer facing one another; two long-waisted animals with tails set at 90 degrees; two shark-like fish; a human with outstretched arms; and a creature with the body of a deer and the head of a human, which is emphasized by outline engraving. The right side of this section features many geometric patterns, most of which consist of lozenge shapes rendered with deep outline chipping. Many of the lozenge shapes are outlined several times, for a mirror-like doubling effect. At the center of the right side, there is a rounded lozenge with an 'X' mark inside, which is reminiscent of a fish with a tail.

The lower part of this section features an assortment of inscriptions, dating as far back as the Silla period, as well as some cross-linked diagonal lines rendered in outline chipping, and animals that look like

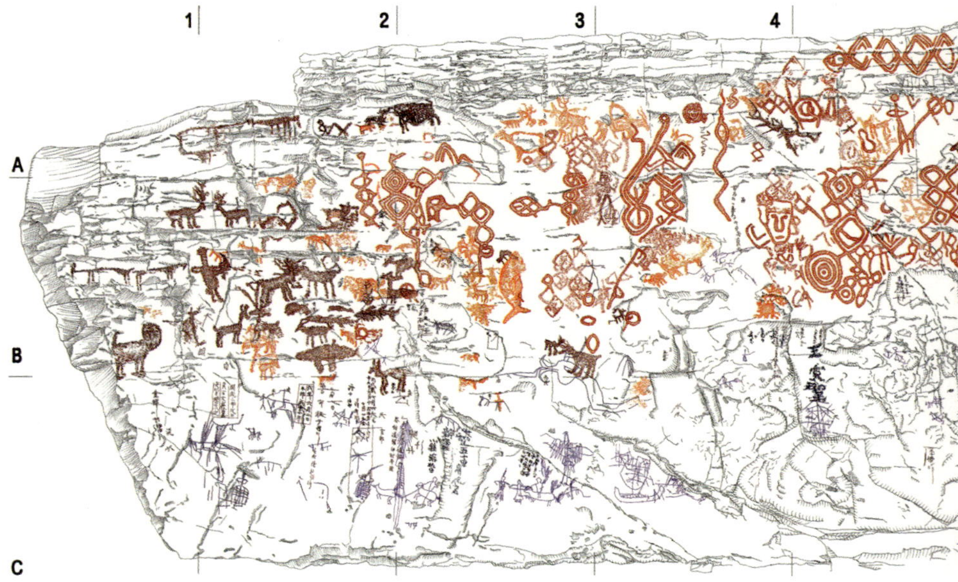

Fig. 3. Diagram of the main rock panel of the Cheonjeon-ri Petroglyphs

horses or deer. Near the left side are two people wearing checkered pants, one of whom is leading a horse. A third human figure is riding a horse in the right part of the section, rather far from the first person. On the right side of the damaged rock is a procession of people, composed of a person on horseback holding a parasol, a person wearing a skirt or kilt, a person on a horseback carrying a quiver, and a spiky-haired person on horseback. The procession is followed by two boats (Fig. 4). Above the procession are several inscriptions, including "Third Year of Gaeseong" (開成三年); "Mu Year" (戊年); "Byong-Sul Year" (丙戌); "Eul-Mi Year" (乙未); and "Gae-Hae Year" (癸亥) (Jeon H. 1999).

② The upper part of the center section features various figures rendered with deep outline chipping, as well as a few figures in silhouette. Most of the silhouette figures were damaged by the outline chipping used to render the later figures. The silhouette figures include a person with his hands on his waist[1], a person with his hands on his head, and a small animal in the upper left part. The upper middle part of this section features an eroded depiction of antlers and the head of a deer, along with

[1] For simplicity, the male pronoun is used unless the gender of the figure is obviously female.

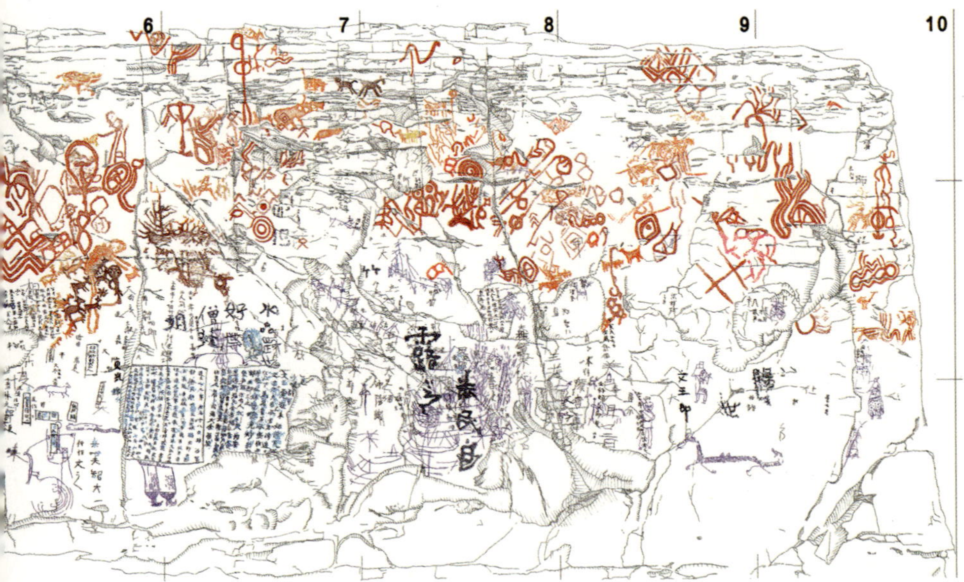

a group of about ten small animals. Near the animals is a flower-like object and a male figure with an exaggerated head that looks like a mask, a long penis between his legs, and one hand resting on his waist.

The deep outlined patterns include an oval shape with a vertical line inside; a series of linked lozenge shapes; a triple concentric circle with three dots inside; a squiggly snake-like line, and more. These geometric patterns are concentrated in the central area (Fig. 5). Near the upper edge of the rock panel are five linked triple lozenges, and the area below this pattern is crowded with different images. There is an oval pattern; a mask-like rectangular human face with a long nose; sets of triple and quintuple concentric circles; a series of double lozenges with two horizontal or vertical lines drawn inside; a group of triple lozenges; a group of linked lozenges; a squiggly pattern; a human figure with a huge head and outstretched arms, and more. The right side has been damaged by weathering, but a few patterns can still be discerned, including small circles, a triple concentric circle without a central point, and some short squiggly lines.

Fig. 4. Diagram of thin-line engravings from left section of the Cheonjeon-ri Petroglyphs

Fig. 5. Lozenge-shaped patterns in the center section of the Cheonjeon-ri Petroglyphs

The entire lower left part of the center section has been lost, so it is not known what, if any, images might have been inscribed there. The lower middle part of the section contains several thin-line engravings: a horse; a dragon-like animal with a scaled body, long tail, and no legs; an animal with a long neck or body; a quintuple semicircle; and a number of irregular straight lines, mingled with short inscriptions. On the lower right side of the section is the upper body of a horse that has been marked with long inscriptions. Below the inscriptions, there is a pair of human legs wearing wide-legged pants and shoes with pointed toes; the hem of the figure's jacket also bears some inscriptions (Fig. 6).

Fig. 6. Diagram of thin-line engravings from center section of the Cheonjeon-ri Petroglyphs

Fig. 7. Inscriptions "Wonmyeong" (原銘, left) and "Chumyeong" (追銘, right) from the Cheonjeon-ri Petroglyphs

Two of the most prominent inscriptions are known as "Wonmyeong" (原銘) and "Chumyeong" (追銘) (Fig. 7). Other scattered inscriptions include

Fig. 8. Diagram of thin-line engravings from right section of the Cheonjeon-ri Petroglyphs

dates and the names of monks and members of the Hwarang, an elite youth group of Silla. There are other inscriptions with unclear meanings.
③ The upper part of the right section of the main rock panel features some traces of chipped silhouettes. Some of the figures that are still recognizable include two small dog-like animals and a number of other animals represented in pairs. Figures in deep outline in the upper part include a quadruple spiral pattern; patterns of curves or waves; a group of small circles; five linked lozenges; double lozenges with a central dot or horizontal line inside; and a vertical oval pierced by an arrow.

The lower left part of this section includes deep-outlined figures as well as inscriptions. The figures include a bow-shaped weapon, four small ship-shaped objects, and various vertical and horizontal lines. Other figures rendered with thin outline incisions include a person in a square frame, another person wearing a skirt, a long dragon-like animal, and more (Fig. 8). Among these engraved geometric patterns and figurative images are numerous inscriptions, including "Eul-Myo Year" (乙卯).

(B) Surface B is a rock panel located 1.0 meter to the right of the main rock panel (Surface A). Surface B includes four or five deer-shaped animals represented in silhouette, but they are extremely weathered and thus difficult to discern (Fig. 9).

(C) Surface C is 2.0 meter away from the protective iron railing of Surface B. The chipped surface is about 1.0 meter high and 2.0 meter long, but the figures are severely damaged and barely identifiable. Based on the analyses of the adjacent rock panels, the figures are probably land animals (Fig. 10).

Fig. 9. Diagram of Surface B of the Cheonjeon-ri Petroglyphs

Fig. 10. Diagram of Surface C of the Cheonjeon-ri Petroglyphs

(D) Surface D, located to the right of Surface C, includes some deep-outline shapes that are still clear, including a double concentric circle, wave patterns, and a group of circles. Some traces of older figures can be partially discerned beneath the shapes, but they are difficult to identify because of weathering. Surface D-2 is located to the upper right of Surface D-1, and it features a series of double lozenges, as well as a double concentric circle that is barely noticeable.

3) Current Research

(1) Petroglyphs

The Cheonjeon-ri petroglyphs are a representative example of prehistoric rock carvings, which helped introduce Korean scholars to the new field of petroglyphs. The carvings date from the sixth to ninth century of the Silla Kingdom. Featuring diverse figures and images, as well as numerous inscriptions of Chinese characters, the Cheonjeon-ri petroglyphs have understandably attracted considerable attention from historians, archeologists, and art historians.

The Cheonjeon-ri petroglyphs were first discovered on December 25, 1970 by a group of researchers from Dongguk University who specialized in Buddhist artifacts (Mun 1973; Hwang and Mun 1984). Since there was no known research relating to the petroglyphs, subsequent studies have focused on simply categorizing the carving techniques and describing the contents of the rock art. Researchers were immediately interested in the inscriptions, which contained information about the royal family of Silla (Yi M. 1992). The petroglyphs have proven to be quite a challenge for researchers, because they feature numerous overlapping figures and images, both realistic and fantastic, from multiple time periods. Thus, scholars have had a difficult time separating and interpreting the various carvings and cultural meanings of the Cheonjeon-ri petroglyphs.

The petroglyphs have been roughly dated between the mid-Neolithic period and the Bronze Age. They include geometric patterns, which are commonly associated with the Bronze Age, but another theory suggests that the patterns were carved by the "patterned pottery" tribes of the Neolithic period (Mun 1973; Hwang and Mun 1984). This theory attracted much academic interest, because there has been little research on Neolithic relics, but its validity has not been confirmed. Another theory suggests that the geometric patterns may be a symbolic rendering of genitalia (Kim Y. 1983), but as yet there is no hard evidence to support this premise.

No substantial research has been conducted on the petroglyphs in quite some time. About ten years after the early research, it was suggested that the geometric patterns were meant to embody divine figures and concepts associated with the agricultural society of the Bronze Age, such as the sun, the growth and maturation of crops, and female genitalia (Song H. 1993). Scholars are quite interested in this theory, because it would mean that the geometric patterns were likely involved in some type of ritual of a religious nature.

Jeong Dong-chan (1996) has proposed that the petroglyphs can be roughly divided into two categories: animal patterns carved by silhouette chipping, and geometric patterns carved with deep outline chipping. According to Jeong, the animal patterns likely represent wishes for abundant game and successful hunting, and thus represent the work of a society of hunters, while the geometric patterns are more emblematic of a society engaged in agriculture and fishing, in that they are likely related to water or a shaman's ritual rain. The issue of the authority or source behind the symbolic meaning of the petroglyphs requires further discussion, but the attempt to associate the petroglyphs with actual behaviors related to religious rituals and a subsistence economy is a meaningful achievement worthy of academic recognition.

Other researchers have suggested that the geometric patterns symbolize the female genitalia or animal mating, and thus represent the wishes of prehistoric people related to "procreation and rebirth" (Jang M. 2001). Another recent study asserts that the petroglyphs on the upper part of the rock panel and the inscriptions on the lower part may have been made around the same time (Kim T. 2003), but there is currently no hard evidence to support this hypothesis.

(2) Thin-line Engravings and Inscriptions

The Cheonjeon-ri petroglyphs include numerous engravings made with very thin lines, mostly located on the lower part of the rock panel. The thin-line engravings, of varying size and composition, are widely scattered on the rock, and many of them are damaged or incomplete. Due to such factors, there is not yet a detailed understanding or interpretation of these engravings. The first comprehensive report on the thin-line engravings was published in 1984 by Hwang and Mun, who surmised that the engravings were made by people from the Sahwuebu (沙喙部) district of Silla, which is mentioned in both the earlier and later inscriptions. If this hypothesis is correct, the engravings were made in the sixth century, but it has been generally agreed that other inscriptions on the lower part of the rock panel range in date from the sixth to the ninth century. Given this range, it is somewhat problematic to assert that all of the thin-line engravings were made by the same group of people during a single time period. Another report concluded that the engravings ranged from the Iron Age to the Unified Silla period (Yim C. 1984), but there is thus far no other evidence to shed light on the thin-line engravings of the Cheonjeon-ri petroglyphs.

In the aforementioned 1984 report, the authors argued that the scaled creature among the petroglyphs was a dragon, based on a comparison to the *cheongnyong* (blue dragon) and *baekho* (white tiger) from the Goguryeo Yaksuri mural painting. Furthermore, they claimed that the procession of people might represent ambassadors from Japan or Arabian merchants (Hwang & Mun 1984). The latter idea attracted some interest among scholars, since there is currently no conclusive evidence of cultural exchange between Silla and Japan, but it remains purely speculative at this stage.

In 1999, Jeon published a study comparing the thin-line renderings of horses and men on horseback to engravings on Silla pottery excavated in Gyeongju, clay figurines from the Silla period, and contemporaneous relics from ancient burial mounds in Japan (Jeon H. 1999). Based on this comparison, Jeon asserted that the thin-line engravings of the Cheonjeon-ri petroglyphs were made in the fifth or sixth century. This study has assumed a place of significance in the research history of the field, because of its extensive research on the social circumstances and cultural productions of the time.

In 2003, an important joint study on the Cheonjeon-ri petroglyphs was published by the Ulsan Metropolitan City and Korean Academy of Petroglyphs, including the introduction of thin-line engravings and inscriptions that had not been included in the 1984 report. The 2003 report emphasized that the thin-line engravings should be closely compared with decorative line engravings from Japanese tumulus, and suggested that the high point of the petroglyph's production might be the sixth century. This suggestion had been evaluated as another achievement in the study of the thin-line engravings of the Cheonjeon-ri petroglyphs.

The two primary inscriptions on the rock panel are known as "Wonmyeong" (原銘) and "Chumyeong" (追銘), and they document that the Silla royal family twice visited "Seoseokgok" (書石谷, "valley of the lettered stone"), in 525 and 539. For obvious reasons, these two inscriptions have attracted much interest from both Korean and international scholars, but their exact meaning has not yet been deciphered. The vocabulary and sentence structure used during the Silla period are notoriously difficult to interpret, so there had been some disagreement about the details of the visit, including who participated, their titles and occupations, and the purpose of the journey (Lee M. 1993; Kang J. 1999). This information was partially explained by the discovery of the Uljin Bongpyeong Silla Stele (524), as well as by studies on the Yeongil Naengsuri Silla Stele (503). These two steles were found to contain essential information about the time of King Galmun of Silla, thereby helping to settle some disputes about the names, titles, and positions of the participants.

The exact purpose of the royal family's two visits to Seoseokgok is still not known. If they were merely traveling for pleasure, then there would seemingly have been no cause to carve the inscriptions. Such inscriptions were often made to commemorate a ritual, but in such cases, the inscription typically describes the ritual in full detail. Thus far, researchers have been unable to solve this mystery (Jeon H. 1999).

Analysis of the Cheonjeon-ri Petroglyphs

1) Thin-line Engravings

(1) Comparison to Relevant Korean and Japanese Artifacts

All of the aforementioned research on the Cheonjeon-ri petroglyphs indicates that the thin-line engravings were made by people of Silla. However, the cultural or religious meaning of the petroglyphs remains unknown, and there have not been any in-depth attempts to investigate this issue. The thin-line engravings of the Cheonjeon-ri petroglyphs are crucial artifacts for our understanding of ancient Korea, and their meaning may be somewhat elucidated by a comparison with other relics.

First, some thin-line engravings of hoofed animals have been found on pottery vessels excavated from the former territory of Silla. For example, a jar excavated from Samgwang-ri, Ulsan features thin-line engravings of horses and deer, and similar engravings have also been found on a plate and lid from unidentified sites. The jar from Samgwang-ri is covered with thin-line engravings of both horses and deer on its shoulder and body. Notably, the horses have three or four incised lines emerging vertically from their hips (Fig. 11), and similar lines can be seen in the Cheonjeon-ri petroglyphs on the hips of the horse with a rider in the procession of people.

A plate excavated from Gyerim-ro tomb 47 in Gyeongju is decorated with thin-line engravings of various animals (e.g. dragons, birds, turtles) and triangle patterns that are believed to represent mountains. The bird has its wings outstretched as if about to fly, and the turtle has its legs spread in a similar pose. The head of the dragon is conspicuously emphasized, and parallel lines were drawn on its body and legs, apparently to represent scales (Fig. 12).

Another intriguing example is an incised earthenware jar from Silla, now

Fig. 11. Silla jar with animal engravings from Samgwang-ri, Ulju-gun, Ulsan

Fig. 12. Silla plate with animal engravings from Gyerimro tomb 47, Gyeongju

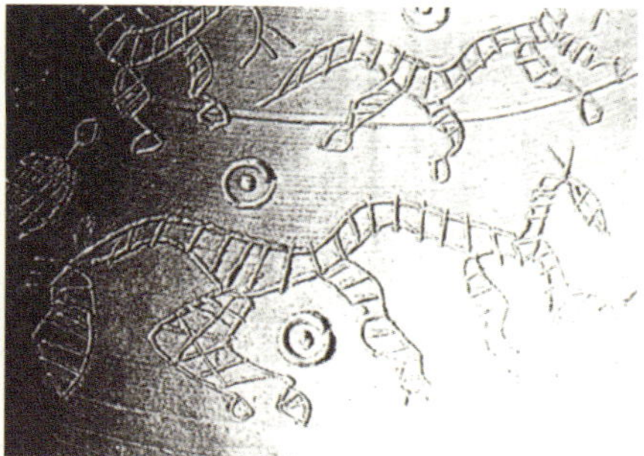

Figs. 13-1, 13-2. Silla jar with animal engravings, excavated from unknown site (Kyoto National Museum, Japan)

Fig. 14. Silla clay figurine of warrior on horseback, excavated from unknown site (Gyeongju National Museum)

Fig. 15. Silla clay figurine of warrior on horseback, excavated from unknown site in Gimhae (Gyeongju National Museum)

Fig. 16. Silla earthenware fragments with dragon decoration, from Hwangnam-dong in Gyeongju, (Gyeongju National Museum)

Fig. 17. Silla clay figures of birds, from unknown site in Gyeongju (Gyeongju National Museum)

housed at the Kyoto National Museum in Japan (NMOK 1997). The shoulder and body of the jar are engraved with images of diverse animals, including dragons, turtles, and fish (Fig. 13). Of greatest interest are the images of dragons, the primary motif of the design, which are engraved in two rows around the shoulder, surrounded by various circles that may symbolize the sun. The top row features a bird flying towards the sun, as well as some reptiles that resemble frogs. There are more birds and dragons in the second row on the lower shoulder. The dragons are rendered in particular detail, with a long, serpentine body, four legs, two horns on the head, and lines drawn to represent scales. The overall style is similar to that of dragons from the Cheonjeon-ri petroglyphs.

Another type of artifact worthy of attention are the Silla clay figurines from the fifth or sixth century, excavated in the area of Gyeongju, particularly the figure of a warrior on horseback excavated from an unidentified location, now held by the National Museum of Korea (Fig. 14). The figurine has a peaked hat that resembles the hat worn by the man on horseback with a parasol from the thin-line engravings of the Cheonjeon-ri petroglyphs. Another example of a Silla rendering of a man on horseback with a parasol can be found on an earthenware vessel excavated in Gimhae (Fig. 15, GNM 1997). The top of the helmet stands straight up, similar to the style of peaked hat from the thin-line engravings of the Cheonjeon-ri petroglyphs. Other analogous elements include the pattern on the horseman's armor, the narrow cut of the clothing, and the use of horizontal and vertical lines.

Another interesting comparison can be made with some earthenware fragments found in Hwangnam-dong, Gyeongju, which feature a dragon marked with parallel lines (Fig. 16), similar to the style on the aforementioned earthenware jar. The long, serpentine shape of the dragon is also comparable to the one in the thin-line engravings of the petroglyphs. Also found in Gyeongju were numerous Silla clay figures of birds, which also resemble bird images

Fig. 18. Detail of bronze bell from Kamoiwakura tomb No. 35 in Shimane, Japan

Fig. 19. Fragment of *haniwa* from Choshiyama tomb in Kyoto, Japan

Fig. 20a. Sue ware jar from Wanagaura Tomb W1 in Fukuoka, Japan

Fig. 20b. Detail of Fig. 20a

Fig. 21. Line-engraving of horseman from wall of Takaida tombs No. 2-3 in Osaka, Japan

Fig. 22. Line-engraving of ship from wall of Takaida tombs No. 2-12 in Osaka, Japan

from the Cheonjeon-ri petroglyphs (Fig. 17).

Also of interest are some Bronze-Age bells for rituals, many of which have been excavated from tomb sites in the Shimane (島根縣) area in Japan. For example, bronze bells recently excavated from the Kamoiwakura site (加茂岩倉かもいわくら) in Japan are decorated with engravings of animals (SEC 1997). Of these, bronze bells found in Kamoiwakura tombs No. 23 and No. 35 feature engraved images of animals with a simple outline for the body and four lines representing the legs (Fig. 18), which resemble the horses in the procession from the Cheonjeon-ri petroglyphs.

Other images similar to the thin-line engravings of the Cheonjeon-ri petroglyphs can be found on Sue ware (須惠器 すえき, ritual pottery); *haniwa* (埴輪 はにわ, clay funerary figures); and other pottery from the period of the aforementioned ancient Japanese tombs. For example, excavations at the fifth-century tomb Choshiyama (銚子山古墳 ちょうしやまこふん) in Kyoto uncovered a cylindrical *haniwa* (圓筒植輪) featuring a line-carving of a dragon. Also, a Sue ware vessel excavated from the sixth-century tomb Iwanagaura (岩長浦いわながうら) IW1 in Fukuoka bears a line-engraving of a horseman (NHAFM 1993). Another line-carving of a horseman can be found in tomb No. 2-3 of the Takaida tomb site (高井田横穴群) in Osaka, and tomb No. 2-12 of that site feature line-carvings of a ship with sails (WUMRI 1978). The tombs and images are believed to date from the mid-sixth to seventh century. Another line-carved dragon appears on the surface of the cylindrical *haniwa* that was excavated from the Choshiyama tomb in Kyoto. The body of the dragon figure is rendered with an outlined shape (Fig. 19), and the thin-line dragon figure on the Cheonjeon-ri petroglyphs is engraved in a similar way.

On the Sue ware jar excavated from Iwanagaura tomb IW1 in Fukuoka, the horse and rider are expressed very simply, with few lines (Figs. 20a, b). The style is similar to that of the horseman figure on the far right side of the Cheonjeon-ri petroglyphs, as well as the figures from the procession. Notably, in the procession, the second figure from the right is holding the reins of the horse, much like the horseman image from Takaida tomb No. 2-3 (Fig. 21). As for the ship carved on the walls of Takaida tomb No. 2-12 (Fig. 22), the shape of the mast is different from its counterpart in the procession from the petroglyphs, but the shape of the outline is similar, and both images feature a simple line-rendering of a person on the deck of the ship holding the wheel.

(2) Significance of Thin-line Engravings of the Cheonjeon-ri Petroglyphs

The thin-line engravings on the left side of the petroglyphs include a walking figure with wide pants, a horseman, two processions, and a ship (Fig. 4). The walking figure has distinct features, including eyes, nose, mouth, and hair that stands up. At some distance from this figure are a pair of processions and the ship. A human figure in the middle of the procession also has upright hair, akin to bird feathers, as well as wide pants, which are called *daeguko* (大口袴). Behind the procession are a horseman leading his horse by the reins. Part of the rock behind the horseman has broken away, and it is unknown whether there was once a petroglyph on that area.

Between these rock surfaces is another procession, led by bearded horseman carrying a parasol and wearing a striped coat and a tall peaked-hat. Behind him is a person wearing a kilt, followed by two more horsemen facing to the left. The first of the two horsemen is holding the reins and what seems to be a quiver, with several lines emerging from it to represent arrows. The other horseman has a triangular head with several oblique lines reminiscent of feathers. All of the horses and people are rendered with very simple outlines.

Behind the procession are two ships with sails. The one on the left is larger and its sail is rendered in more detail. The ship on the right, which seems to be sailing out at sea, has an outlined sail and a figure standing in the prow of the ship. There is a diagonal line engraved to the right side above the procession and ships, and according to the comprehensive report from 1984, this area once contained more simple engravings of horsemen, but they are no longer discernible. The report indicates that, although the image was rendered with simple lines, the motion and physical features of the horse and its motion were clearly expressed.

It is not clear whether the processions and the ships are meant to be linked or if they are all separate images. However, the two are very similar in terms of the engraving technique and level of expression (Jeon H. 1999). Moreover, there is a certain consistency in the types of figures included in the processions. Thus, it seems likely that the three sets of images can be combined to formulate a single narrative of the people of Silla welcoming visitors from a foreign land, perhaps Japan or Gaya, and then leading them in a parade. If this interpretation is correct, the first man in the wide pants must be a delegate from the Silla people, leading the procession. The lines emerging from his head are probably meant to depict actual feathers, which were a favored decoration in Silla. The man carrying the parasol would then likely be the leader of the foreigners visiting Silla, and his clothing might actually be armor. These assumptions are supported by the fact that images of horsemen in armor have frequently been found on *haniwa* excavated from ancient Japanese tombs of the Kobun period. Armored horsemen can also be seen on Gaya pottery of the fifth and sixth century. The human figures in the processions are certainly worthy of the considerable scholarly attention they have drawn, given that they may represent evidence of exchange between Silla and Gaya or Japan.

In the lower right section of the Cheonjeon-ri petroglyphs, there are engravings of a person and a horse, although parts of these two images have been chipped off by the inscriptions (Fig. 6). Other images in this area include the legs of a human body and several animals, including a dragon, a long-necked animal, birds, and fish. To the left are four animals that are presumed to be horses. Next to them is a human figure that is expressed much differently from people in the previously discussed procession (Fig. 8). At 28 cm in height, this figure is larger and taller than the others, and he is facing directly outward. His face is rectangular with the eyes, nose, and mouth expressed in detail. Ten lines are carved on his chest, which seem to represent his fingers, and he wears narrow trousers that are tied at the ends.

The human figure who are carved in this section, but the upper half of the body has been effaced by the inscriptions, so only the legs remain. The figure is wearing wide patterned pants and shoes with pointed toes, the outfit of a Silla aristocrat. Above him is a horse, but only the upper half of the body remains, because of the inscriptions. From what remains of this figure, we can tell that it was originally engraved in great detail, more so than the figures in the procession in the left section of the petroglyphs. Also, the techniques used to depict this figure and the horse are more advanced than those used to render the procession figures. These differences have drawn considerable interest from various scholars.

The inscription that covers the upper part of the human figure and lower part of the horse is called "Chumyeong" (追銘), and it is estimated to have been made in 539. The placement of this inscription is quite curious, because it completely destroyed the upper half of the male figure. Given that there was ample space available on other parts of the rock, it almost seems as if the Chumyeong inscription was deliberately rendered in its location, despite the inevitable damage that would be caused to the figure. Such deliberate placement would indicate that the inscription held some special meaning or significance that superceded the other engravings, which in turn would seem to signify that the Cheonjeon-ri petroglyphs were more than just a simple scribbling board. The damage to the engravings and the new inscriptions may have held some political or religious significance, but the nature of that significance remains a mystery. Various researchers have endeavored to decode the Wonmyeong and Chumyeong inscriptions, and they have thus far been able to produce the following interpretation (Gang J. 1999).

Table 1. Wonmyeong

XII	XI	X	XI	IX	VII	VI	V	IV	III	II	I	
作	眞	榮	悉	食	鄒	幷		之	文	沙	乙	1
書		知	淂	多	女?	遊	以	古	王	喙	巳	2
人	智	智	斯	煞	郎	友	下	谷	覓	部		3
慕	沙	壹	智	作	王	妹	爲	无	遊	葛		4
	干	吉	大	功	之	麗	名	名	來			5
尒	支	干	舍	人		德	書	谷	始			6
智	妻	支	帝	尒		光	石	善	淂			7
大	阿	妻	智	利		妙	谷	石	見			8
舍	兮	居	作	夫		於	字	淂	谷			9
帝	牟	知	食	智		史	作	造				10
智	弘	尸	人	奈			之					11
	夫	奚		□								12
	人	夫										13
		人										14

In the Eulsa year (乙巳, 525), King Galmun from the district of Sahoebu (沙喙部) and his entourage came to the valley for pleasure. It was their first time visiting this place. The valley was very old, but it did not have a name. Therefore, we named the valley "Seoseokgok" (書石谷, "valley of the lettered stone"). We inscribed the name on a stone, along with some other phrases, and placed the stone. Galmun was accompanied by his sister, Queen Eosachunyeorang (於史鄒女郎).

The officers who conducted the ritual consisted of Iribuji (尒利夫知), Daenama (大奈麻), and Sildeuksaji Daesajaeji (悉淂斯知 大舍

帝智). The cooks were Lady Geojisihye (居知尸奚), wife of Youngjiji Ilgilganji (榮知智 一吉干支) and Lady Ahyemohong (阿兮牟弘), wife of Jinyukji Saganji (眞肉知 沙干支). The one who composed this inscription was Momoiji Daesajaeji (慕慕尒智 大舍帝智).

Table 2. Chumyeong

XII	XI	X	XI	IX	VII	VI	V	IV	III	II	I	
一		居	作	支	叱	愛	妹	王	部	過		1
利	知	伐	功	妃	見	自	王	共	徙	去		2
等	淝	干	臣	徙	來	思	過	遊	夫	乙		3
次	珎	支	喙	夫	谷	己	人	來	知	巳		4
夫	干	礼	部	知	此	未	丁	以	葛	年		5
人	支	臣	知	王	時	年	巳	後	文	六		6
居	婦	丁	礼	子	共	七	年	□	王	月		7
礼	阿	乙	夫	郎	三	月	王	□	妹	十		8
次	兮	尒	知	深	來	三	過	八	於	八		9
□	牟	知	沙	□		日	去	□	史	日		10
干	呼	奈	干	夫	另	其	其	年	鄒	昧		11
支	夫	麻	支	知	卽	王	王	過	女	沙		12
婦	人	作	□	共	知	与	妃	去	郎	喙		13
沙	尒	食	泊	來	太	妹	只	妹				14
爻	夫	人	六	此	王	共	沒	王				15
功	知	眞	知	時	妃	見	尸	考				16
夫	居			□	夫	書	兮					17
人	伐				乞	石	妃					18
分	干											19
共	支											20
作	婦											21
之												22

On the eighteenth day of the sixth lunar month of the Eulsa year (乙巳, 525), King Galmun (徙夫知葛文王) of Sahoebu (沙喙部) and his sister, Queen Eosachunyeorang (於史鄒女郎), visited here. Eight years later, the sister of the king has become a person of the dead. Then in the Jeongsa year (丁巳, 536), the king died. Because the queen missed him so much, on the third day of the seventh month of the Kimi year (己未, 539), she came to see the Seoseok (書石, lettered stone) in the valley where the king and his sister used to come. A total of three people came to visit: Great Queen Yeongjeukji (另卽知), Bugeuljibi (夫乞支妃), and Prince Sabuji Sim□buji (徙夫知 深□夫知).

The one who composed was Jiraebuji Saganji (知礼夫知 沙干支), □bakyukji (□泊六知) Geobeolganji (居伐干支) from Whaebu (喙部). The one who conducted the ritual was Jeongyeuliji Nama (丁乙尒知 奈麻). The cooks were Lady Ahaemoho (阿兮牟呼), wife of Jinyukji Pajinganji (眞肉知 波珍干支); Lady Ilrideungcha (一利等次), wife of Yibuji Geobeolganji (尒夫知 居伐干支); and Lady Sahyogong (沙爻功), wife of Georaecha□ganji (居礼次 □干支).

Thus, the Wonmyeong inscription was apparently carved in the Eulsa year (乙巳, 525, 12th year of King Beopheung's reign), when King Galmun (徙夫知葛文王) and his company visited the valley for the first time and named the valley Seoseokgok. King Galmun, who was the father of King Jinheung (眞興王), died in the Jeongsa year (丁巳, 536). Then in the Kimi year (己未, 539, 26th year of King Beopheung's reign), King Galmun's queen and her group came to the valley, and the Chumyeong inscription was engraved. The latter inscription was rendered over the top of the figures of the man in wide pants and the horse, significantly damaging them. Based on the content of the inscriptions, those figures were probably relatively intact at the time, so it is not clear why the queen and her group chose to damage them by making their inscription.

Several of the names listed among the second party to visit the valley in the Kimi year are worthy of particular interest, including Great Queen Yeongjeukji (另卽知太王妃), Prince Sabuji (徙夫知王子郞), and Bugeuljibi (夫乞支妃), i.e. Lady Bodo (保刀夫人), the wife of King Beopheung (法興王). Prince Sabuji may refer to King Jinheung, King Galmun's son; if so, then the most important members of the Silla royal family participated in the second journey to visit Seoseokgok.

Furthermore, after King Beopheung died, Lady Bodo, as the mother of King Jinheung, served as queen regent for one year while her young son came of age. Clearly, the visit to Seoseokgok in the Kimi year was an extremely important event for the Silla royal family.

Other historical evidence has shown that the period between 525 and 539 was one of relative turmoil for the Silla royal family, due to some severe conflicts regarding the official recognition of Buddhism. Specifically, during this time, the royal family under King Beopheung strongly supported the official authorization of Buddhism as the state ideology. This period also saw some major political changes related to the founding of the Sangdaedeung (上大等) system and the bestowal of the official title of "Great" (太王) to the kings. As indicated by the Seoseok inscription Eulmyomyeong (乙卯銘, 535, 22nd year of King Beopheung's reign), when King Beopheung came to the throne, he was called the "Divine King Beopheung" (聖法興大王). Based on this interpretation, the person who carved the second inscription and damaged the upper body of the engraving of the Silla aristocrat apparently was from Sahoebu and was a member of the Silla royal family. Perhaps the figure whose upper part was damaged was thought to represent a person or group who protested the king's sovereignty or the religious ideas promoted during the reign of King Beopheung. (Jeon H. 1999).

The petroglyphs also feature three animals that are presumed to be dragons. To the left of the Chumyeong inscription is a dragon standing upward, next to what appears to be a lake with a rippling or boiling surface. Above these images is another dragon in a prostrate position, as if it were crawling. This dragon has a long coiled tail, like a snake, and some dot patterns on the body. The third dragon, located in the right section of the petroglyphs, seems to be moving forward with its head raised. It has a very long body (57 cm) covered with scales, and its tail is curled up in an "S" shape (Fig. 23).

The first dragon is rendered very simply, without details such as scales. But, as suggested in the 1984 report, the fact that the dragon is depicted next to a body of water would seem to suggest that it is a wyvern, a type of dragon commonly associated with the sea (Hwang and Mun 1984). The head of the second dragon resembles that of a tiger, and the dots on its body may represent reptilian scales. If so, this image is reminiscent of the White Tiger from the Goguryeo tomb murals in Pyeongyang. Notably, the third dragon has the same

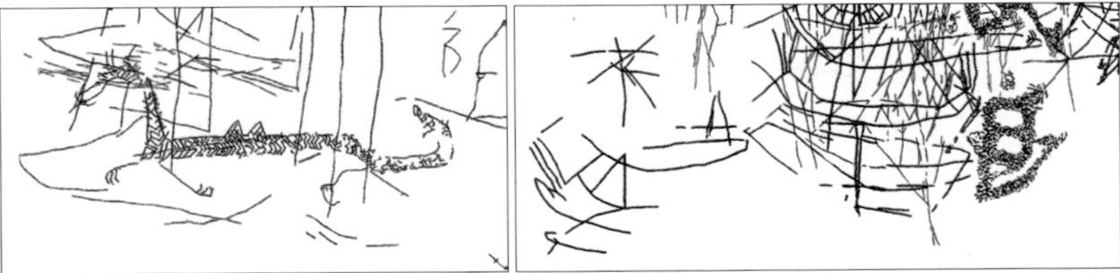

Fig. 23. Thin-line engraving of a dragon from the Cheonjeon-ri Petroglyphs

Fig. 24. Thin-line engraving of birds from the Cheonjeon-ri Petroglyphs

pose as the Blue Dragon and White Tiger from the second-phase Goguryeo tomb murals. Hence, like the Blue Tiger on the Yaksuri tomb or Suryeopchong (Tomb of Hunters), this dragon may represent the Blue Dragon (Jeon H. 1999).

To understand why the petroglyphs would contain three very different forms of the same mythical creature, we must consider the significance of dragons in Silla culture. Dragons appear in the Silla myths of Bak Hyeokgeose (朴赫居世), Alyeong (閼英), and Algi (閼智) (*Samgukyusa* 三國遺事), where they embody the secret power that generates great kings and queens. With the official recognition of Buddhism, dragons became symbols of power for protecting the nation, with the Buddha's blessing. There is a myth about a Chinese ambassador who secretly tried to capture the dragon protector of Silla, but failed. According to another story, King Munmu (文武王), thirtieth king of Silla, became the dragon guardian of Silla after his death (*Samgukyusa* 三國遺事). Most of the known visitors to Seoseok were members of the royal family, monks, and ambassadors, all of whom would likely have been loyal to the royal family and have wished for the well-being of the country. Some of them also would surely have appreciated the role of dragons in the history of Silla. Accordingly, we can assume that some of these visitors engraved dragons to express their wish for the peace of the royal family and the country.

According to the comprehensive report, the thin-line engravings include images of five birds (Hwang and Mun 1984). One bird has three lines seemingly representing its crest, while another has three lines denoting the plumage of its tail. One bird, which seems to represent a bird on a pole, has various detailed features, including eyes; a beak; two long, slightly stretched legs; and a body shaped like a ship (Fig. 24). There is also one image of a fish, carved with a simple 4.3 centimetre line.

The birds were probably engraved for religious reasons, because ancient societies considered birds to be mediators between heaven and earth. Birds, as heralds of the gods, were believed to be capable of carrying people's thoughts to the gods and also of taking the souls of the deceased into the afterlife. These beliefs are well documented in the Chinese historical text *Samgukji* (三國志, Chronicles of Three Kingdoms), which states that people used large feathers to make funeral shrouds. The same record confirms that the people of Goguryeo often wore feathers on their head, a practice that was related to their religious belief. Further proof of the religious significance of birds can be found in the myth of the founding of the Silla nation. According to the myth, a child named Alyoung was born to a dragon, and the baby had the beak of a bird, but the beak later washed away when Alyoung was bathed in Bukcheon Stream in Gyeongju (*Samgukyusa*). Alyoung later married Bak Hyeokgeose, founder of Silla. Thus, the Silla people believed that the mother of their country was a bird incarnate. As such, the bird images in the Cheonjeon-ri petroglyphs reflect the deep religious reverence that the Silla people held for birds.

There are also images of a tall human figure and a procession (Fig. 25), which was engraved vertically as if viewed from the back side. The horseman and horse in the front were rendered almost in profile. The man following them is wearing a hat and has what looks like a sword in his belt. He is wearing wide pants decorated with a pattern of circles and triangles, implying that he is an aristocrat. It is not clear if he is riding a horse or not. Two more people follow this man, both engraved from the rear. These images are peculiar because of the rear perspective. Attempting to express a rear perspective in a rock carving is very difficult, and requires advanced skills of engraving. Thus, further studies should focus on this particular set of images, in terms of its possible time period and its relation to Silla painting techniques.

Fig. 25. Thin-line engravings of a tall human figure from the Cheonjeon-ri Petroglyphs

The thin-line engravings of the petroglyphs,

which are presumed to have been made around the fifth or sixth century, provide valuable information about Silla society. In this case, the images of the human figure and the procession shed light on the costume style of the time, the social structure, and even the artistic technique, although our understanding remains limited. Also, the images of dragons, birds, and other creatures allow us to understand religious trends of the time. In particular, the thin-line engravings confirm the existence of dragons in the founding myth of Silla, as well as the religious faith in birds.

As mentioned, however, the content of the thin-line engravings of the petroglyphs is so simple that it is extremely difficult to determine its relation to the surrounding inscriptions. With the methods currently available to us, the best we can do is to make some rather reckless guesses. Further research related to the thin-line engravings must be conducted, but more importantly, new scientific methods for evaluating the petroglyphs are required.

Inscriptions

The Cheonjeon-ri petroglyphs include various inscriptions, most of which were carved from the sixth to the ninth century (Yi M. 1992; Kang J. 1999). The inscriptions can be interpreted according to the social classes and governmental positions of the people named in the carvings.

Some of the people named in the inscriptions are known to have held very high social positions, including members of the Silla royal family. Most notably, the Wonmyeong inscription was made in the Eulsa year (乙巳, 525, 12th year of King Beopheung's reign), and the Chumyeong inscription was made in the Gimi year (己未, 539, 26th year of King Beopheung's reign). Both inscriptions include the names of King Galmun of Sahoebu (King Beopheung's younger brother, King Yipjong-galmun), his wife, his son (King Jinheung), and King Beopheung's wife, who was both sister-in-law and mother-in-law to King Galmun. The inscriptions also include the names of various aristocrats from Sahoebu, such as Gaesamyung, Eulchukmyung, and Gaehaemyung.[2]

Other inscriptions include the names of Buddhist monks. There are multiple inscriptions dated with *ganji* (干支, years from the Chinese sexagenary calendar cycle). For example, there are inscriptions from the Eulmyo year (乙

[2] Inscription reads: 癸巳銘: 癸巳六月卄二日 沙喙 壹奮…王夫…輩衆大等…部書人小…[下略]);
(乙丑銘: 乙丑年九月中 沙喙部 于西□夫智 彼珍干支妻 夫人阿刀郎女 谷見來時 前立人…[下略]);
(癸亥銘: 癸亥年 二月八日 沙喙 □凌智小舍婦 非德刀 遊行時書)

卯, 535, 22nd year of King Beopheung's reign)[3] and the Gabin year (甲寅, presumed as 534 or 596).[4] Another inscription has no *ganji*, but does include the term "Seungdo" (僧徒).[5]

The type of name that appears most frequently in the inscriptions are the names of members of Hwarangdo (花郎, a powerful group of elite Silla youth) and Nangdo (郎徒, followers of the Hwarang). Most of these inscriptions only include the names, but some also include *ganji*. Several inscriptions with *ganji* were inscribed after Silla unified the three ancient kingdoms of Korea in 668 (e.g. 丙戌銘 [746], 丙申銘 [756]). Some of the names are from members who served in Hwarangdo before the unification, which proves that Seoseok and the Cheonjeon-ri petroglyph rock were well-known to the Hwarang, even before the founding of the Hwarangdo (花郎徒) in the sixth century. Perhaps most interesting is the coincidence of names of Hwarang and names of Buddhist monks. Future research might attempt to explain the significance of the Cheonjeon-ri petroglyphs as chronicles for Silla people of various classes and social positions.

Some studies have claimed that the Cheonjeon-ri petroglyphs must have been a ritual site (Yi W. 1997), while others feel that it was merely a popular area for sightseeing (Ju B. 1997). Supporters of the first hypothesis point to the Bronze-Age geometric patterns among the petroglyphs, the type of which have been commonly associated with rituals. The second hypothesis is supported by the fact that there are no markings or references within the Cheonjeon-ri petroglyphs naming the place as a ritual site. As discussed, the Wonmyeong inscription was made when King Galmun came to the valley for sightseeing, but the later Chumyeong inscription was made by his family when they revisited the place to memorialize him after his death. Thus, even if the Wonmyeong inscription was merely the result of a simple sightseeing visit, the Chumyeong inscription clearly served as commemoration, indicating that the area had some religious significance.

In examining the issue, it is essential to understand why others—such as Buddhist monks, Hwarang, and aristocrats—visited the site. Such visits may

[3] Eulmyo inscription reads: 乙卯年 八月四日 聖法興大王節 道人比丘僧 安及以 沙彌僧 首乃至 居智伐村 衆士□人等 見記.

[4] Gabin inscription reads: 甲寅 大王寺中 安藏 許作.

[5] Seungdo inscription reads: □□王七年僧徒□□, 慕谷僧徒于另□□.

have been only for sightseeing, or for training the mind and body, or perhaps for prayer, given the mystical atmosphere of the site.

Outlook and Future Tasks

Thus far, two major studies have documented the results of the examination of the Cheonjeon-ri petroglyphs, its inscriptions, and thin-line engravings (Hwang and Mun 1984; UMC, KPAI 1997). More recently, a research team from Kukmin University Museum launched "Total Actual Examination" with regards to the petroglyphs, presenting some of their results at an academic symposium (UUM, KPRI, UMBC, 2004). Coming twenty years after the initial report, these efforts are long overdue, and they are most useful and meaningful to the academic community. Of course, trying to conduct research on the petroglyphs based on false information is like trying to construct a building on a foundation of sand. Still, many important research tasks and responsibilities related to the petroglyphs remain for the academic community of today and tomorrow.

First, the results of the detailed examination are far from infallible, to the point where they cannot even be completely trusted. The recent studies have academic significance, but they have failed to objectively track the process of how the images and inscriptions were engraved, mainly because they have relied too much on the early examination results and interpretations. As a result, the petroglyphs have not been adequately categorized. Thus, more detailed examination of the petroglyphs is needed.

Second, further research is required to analyze the petroglyphs on the upper part of the rock panel, particularly in terms of their relation to the surrounding area and other sites with similar petroglyph traditions (Jeon H. 1996). Thus far, the upper petroglyphs have been considered only in the context of Neolithic and Bronze-Age cultures. As such, they are generally classified in simple binary terms, as either animal or abstract patterns, or else as either realistic or symbolic art. However, recent examinations have revealed a much wider variance in the composition of the petroglyphs (UUM, KPRI, UMBC 2004).

In this regard, the petroglyphs must be studied in terms of their association with hunting and farming culture, historical facts, and religious activities. Any theories that emerge must be supported with other evidence, apart from the petroglyphs themselves. The area surrounding Cheonjeon-ri and areas with similar petroglyphs should be examined to explicate any possible relationship among the sites, as well as to seek links between the engravings and contemporaneous culture.

Third, the inscriptions and thin-line engravings of the petroglyphs must be more comprehensively interpreted in light of current knowledge about Silla culture, society, ideology, and religion. This could prove very difficult, given the small number of inscriptions and the overall lack of knowledge about their relation to one another. The two primary inscriptions have now been adequately deciphered, shedding valuable light on our understanding of the Silla royal family. But the interpretation of the other inscriptions is still in the early stages. Furthermore, some of the inscriptions and thin-line engravings are overlapping, indicating some linkage between them. These relations and their meaning must be studied more actively in the future.

Fourth, a more concrete plan for protecting and preserving the petroglyphs must be enacted. The Cheonjeon-ri petroglyphs were designated as a national heritage in the 1971 academic report, and they have been protected since then. However, the site has not been adequately managed, and people can easily access the site without restriction. Parts of the rock have been damaged and eroded by the illicit production of rubbings, some of which have involved foreign substances like mimeograph ink and synthetic resin. If stricter regulations are not implemented and enforced, the petroglyphs could be severely damaged.

Fifth, further studies must be conducted to determine how to protect the surface of the rock from damage caused by weathering and precipitation. In the forty years since the discovery of the rock panel, there have not been any lithographic studies or evaluations to consider how to protect the surface of the rock. Recently, cracks have begun to appear in the rock, perhaps accelerated by more extreme conditions associated with climate change, i.e. alternation between severe drought and torrential rains. Clearly, steps must be taken to prevent further damage caused by weather.

Sixth, future studies should incorporate a more interdisciplinary approach to re-evaluate the petroglyphs. The Cheonjeon-ri petroglyphs may be seen as a comprehensive language of ancient times, and thus should have great relevance for scholars from numerous fields, including religious studies, anthropology, mythology, ancient ecology, geology, and preservation science. Short studies within separate fields typically emphasize only a single area of focus, which leads to one-dimensional research. The implementation of a system of cooperative research would prevent errors caused by misinformation and narrow perspectives. In order to ensure a positive outcome, interdisciplinary research and cooperation should be backed by funding and support from research institutions.

The Cheonjeon-ri petroglyphs contain a variety of different images and inscriptions from various time periods, and further research is required to explicate the relationship between these different elements. They may have cultural and historical meaning beyond our current understanding. More comprehensive, professional, and systemic research perspectives and methods are needed to examine these significant relics. In particular, future research can uncover more concrete information about when, why, and how the petroglyphs were produced, as well as who produced them. My personal hope is that future studies on the petroglyphs adopt a more interdisciplinary approach in order to both broaden and deepen our knowledge of this significant historical site.

BIBLIOGRAPHY

Primary Sources

Samgukji (三國志)

Samguksagi (三國史記)

Samgukyusa (三國遺事)

Reports and Catalogues

Gyeongju National Museum ed. 1997. *Silla tou: Sillain ui sam, geu yeongwonhan hyeonjae,* Seoul: Tongcheon munhwasa.

Hwang, Su-yeong and Mun Myeong-dae. 1984. *Bangudae ambyeok jogak,* Seoul: Dongguk University Press.

Jeong, Sang-bak and Yu Jong-mok. ed. 1986. *Hanguk gubi munhak daegye,* Seongnam: Academy of Korean Studies.

Jeonju National Museum. 1995. *Teukbyeoljeon bada wa chukje: Buan Jukmak-dong chukje yujeok.*

Korean Cultural Heritage Foundation. 2001. *Daegok-daem pyeonip buji nae sigul mit balgul josa yak bogoseo.*

National History and Folk Museum (Japan). ed. 1993. *Sōshoku kofun no sekai,* Asahi Simbun.

National Museum of Korea. ed. 1997. *Hanguk godae ui togi: heuk, yesul, sam gwa jugeum.*

Simane Education Committee. ed. 1997. *Kamo Iwakura iseki: dōtaku no nazo,* Kawaidae Shobo Sinsa.

Ulsan Metropolitan City and Korean Prehistoric Art Research Institute. 2003. *Gukbo je 146 ho Cheonjeon-ri gakseok silcheuk josa bogoseo.*

Wako University Ancient Mound Study Institute. ed. 1978. *Takaida yokoanagun senkokuga.*

Books and Dissertations

Im, Se-gwon. 1994. "Hanguk seonsa sidae amgakhwa ui seonggyeok," Ph.D. Dissertation, Danguk University.

Jang, Myeongsu. 2001. "Hangguk amgakhwa ui munhwasang e daehan yeongu: sinang ui jeongae yangsang eul jungsim euro," Ph.D. Dissertation, Inha University.

Jeong, Dong-chan. 1996. *Sara inneun sinhwa bawi geurim,* Seoul: Hyean.

Kim, Gi-heung. 2000. *Cheonnyeon ui wangguk Silla,* Seoul: Changjak gwa bipyeong.

Kim, Yeol-gyu. 1983. *Hanguk munhaksa, geu hyeongsang gwa haeseok,* Seoul: Tamgudang

Journal Articles

Gang, Jong-hun. 1999. "Ulju Cheonjeon-ri Myeongmun e daehan il gochal," Ulsan University Museum, *Ulsan yeongu* 1.

Jeon, Hotae. 1996. "Ulju Daegok-ri, Cheonjeon-ri amgakhwa," Hanguk ui amgakhwa, Seoul: Hangilsa.

_________. 1999. "Ulju Cheonjeon-ri seoseok seseongakhwa yeongu," *Ulsan yeongu* 1.

Ju, Bo-don. 1997. "Ulju Cheonjeon-ri seoseok myeongmun e daehan il geomto," Seogo Yun Yongjin gyosu jeongnyeon toeim ginyeom nonchong.

Kim, Tae-sik. 2003. "Sinseon ui wangguk dogyo ui sahoe Silla: jeokseok mokkwakbun gwa geu sidae reul jungsim euro," Cultural Heritage Administration of Korea, *Munhwajae* 36.

Mun, Myeong-dae. 1973. "Ulsan ui seonsa sidae amgak byeokhwa," Cultural Heritage Administration of Korea, *Munhwajae* 7.

Sa, Jeong-hi. 2001. "Silla sidae Ulsan jiyeok bulgyo ui milgyo jeok seonggyeok e gwanhan gochal," *Ulsan yeongu* 3.

Song, Hwa-seop. 1993. "Hanbando seonsa sidae gihamun amgakhwa ui yuhyeong gwa seonggyek," *Seonsa wa godae* 5.

University of Ulsan Museum, Korean Petroglyph Research Association, and Ulsan MBC. 2004. *Ulsan Amgakhwa ui jaejomyeong* (conference proceedings).

Yi, U-tae. 1997. "Ulju Cheonjeon-ri seoseok wonmyeong ui jaegeomto," National Institute of Korean History, *Guksagwan nonchong* 78.

Yi, Mun-gi. 1983. "Ulju Cheonjeon-ri seoseok won, chumyeong ui jaegeomto," *Yeoksa gyoyuk nonjip* 4.

_________. 1992. "Ulju Cheonjeon-ri seoseok," Korean Prehistoric Research Institute, *Yeokju hanguk godae geumseongmun,* vol. 2.